Learning Exercises in Astronomy

David A. Pierce
El Camino College

HOLT, RINEHART AND WINSTON

New York   Chicago   San Francisco   Atlanta   Dallas   Montreal   Toronto   London   Sydney

Figures 10-21 and 11-5 are taken from <u>Modern Astronomy</u>
by Ludwig Oster, © 1973 by Holden-Day, Inc.

This collection of questions, problems and exercises is designed to be a learning aid for students of general astronomy. As such it is intended to supplement and reinforce material learned from lectures and the text, and to replace the traditional homework assignments. The responsibility both for completing the exercises and for checking the solutions is given to the student. The instructor needs only to inform the class as to which of the exercises it will be responsible for.

Each chapter is devoted to exercises on a specific subject area in astronomy. Chapter 3 covers atoms and light, for example. The exercises are further subdivided into specific topics (spectra, doppler effect, etc.). Within each chapter the exercises are followed by hints to their solutions, and then by the solutions.

In order to obtain the maximum benefit from this book, the student should first attempt to complete each exercise using lecture and text materials from the course. When the student feels that he has answered the exercise completely, then he should check the solution for instant feedback and knowledge of results. If the student gets stuck or isn't sure what the question means, then the hint should be consulted. If the hint is not a sufficient aid, then he may consult the solution to see how to proceed. If the student wants to learn astronomy and to learn how to solve problems, then the temptation to quickly look at the solution should be resisted until some attempt has been made to complete the exercise.

The instructor should select those exercises which are relevant to the particular course of study, and then answer any questions the students ask as a result of their self study. There is no need for the instructor to collect papers or grade them. The students derive the same benefit, but more quickly, from the instant feedback they obtain from the solutions.

There are too many exercises for the average student to complete during one quarter or one semester. If general astronomy is a one year course, however, there may be enough time for some students to complete all the exercises. The large number of exercises is included to cover the wide range of material that is presented by various instructors in introductory astronomy courses.

Three different kinds of exercises are included:

1. Exercises that ask for specific information or facts, to emphasize important ideas, to reinforce learning, and to test retention.

2. Exercises of a hypothetical nature to enable the student to apply basic principles in contexts different from those in which they are offered in the texts. These questions test the understanding of basic principles.

3. Problems in which formulas and numerical data must be used.

The more difficult exercises are identified by an asterisk (*). There are few exercises that simply ask for a definition. Rather, the exercises are designed so that the astronomical terminology must be understood before the exercise can be properly completed.

Among the hints are included any specific numerical data or formulas that are needed. Hints are not given for all parts of all exercises.

Solutions are listed for each exercise, some with explanation, some without. Where an exercise asks for a number of answers, I have listed only a few representative samples. The metric system of units is emphasized throughout.

The initial motivation for this kind of exercise book came from watching my wife struggle with chemistry problems, for which there were no hints, no explanations, no solutions, and no feedback of results--only lots of frustrations for her and bad feelings toward the entire subject. It is my hope that this book with its special format will help generate in students positive feelings for the subject of astronomy and for their own abilities to learn college level science.

The author would very much appreciate constructive criticisms of this book and suggestions for exercises that might be included in future editions.

TABLE OF CONTENTS

Preface   i

1 ANCIENT ASTRONOMY   1

   Exercises: Early Astronomical Concepts   1
               Astronomical Observations and Phenomena   1
               Important Early Astronomers   3
   Hints   4
   Solutions   5

2 PLANETARY MOTIONS: HISTORICAL AND MODERN   7

   Exercises: Observations and Early Theories   7
               Copernicus, Tycho and Galileo   7
               Kepler: Motions in the Ellipse   8
               Newton: The Laws of Motion and Gravity   9
   Hints   11
   Solutions   13

3 ATOMS AND LIGHT   17

   Exercises: The Bohr Atom   17
               Reflection, Refraction, Dispersion, and the Inverse Square Law   17
               Wavelength, Frequency, Speed and Energy   18
               Spectra   19
               The Doppler Effect   20
   Hints   21
   Solutions   24

4 ASTRONOMICAL INSTRUMENTS   29

   Exercises: Optics and General Characteristics   29
               Observing with Astronomical Instruments   31
               Auxiliary Astronomical Equipment   33
               Locations of Astronomical Instruments   33
   Hints   34
   Solutions   37

5 THE EARTH: A PHYSICAL BODY   44

   Exercises: Properties of the Earth's Interior and Surface   44
               Properties of the Earth's Atmosphere and Beyond   45
               Influence of the Atmosphere on Astronomy   45
   Hints   47
   Solutions   49

6   THE EARTH: A CELESTIAL BODY   54

    Exercises: Orientation and Motions of the Earth   54
                Seasons on the Earth   56
                Astronomical Coordinate Systems and the Celestial Sphere   57
                Time and Date on the Earth   58
    Hints   60
    Solutions   63

7   THE MOON   70

    Exercises: Appearance in the Sky: Phases and Eclipses   70
                Lunar Surface Features and Environment   72
                Lunar Motions, Gravity and the Tides   73
    Hints   75
    Solutions   77

8   THE SOLAR SYSTEM: PLANETS   81

    Exercises: Planetary Observations and Appearances in the Sky   81
                Planetary Surface Environments and Features   83
                General Characteristics   85
                Planetary Formation and Discovery   85
                Planetary Orbits and Motions   86
    Hints   88
    Solutions   92

9   THE SOLAR SYSTEM: ASTEROIDS, COMETS AND METEORITES   102

    Exercises: Asteroids   102
                Comets   102
                Meteors   104
                Meteorites   104
                Meteoroids   105
    Hints   106
    Solutions   108

10 THE SUN   112

    Exercises: General Properties   112
                Appearance of the Sun and Its Influence on the Earth   112
                Surface Features: Spots and Assorted Blemishes   113
                Energy Production and Transfer--The Solar Interior   115
    Hints   116
    Solutions   118

11 GENERAL STELLAR FEATURES   122

    Exercises: Stellar Parallax and Distance   122
                Observed Stellar Motions   122
                Stellar Magnitudes and Colors   124
                Spectral Class and the H-R Diagram   126
    Hints   128
    Solutions   131

12 MULTIPLE STAR SYSTEMS   136

    Exercises: Binary Star Systems   136
                  Star Clusters   137
    Hints   140
    Solutions   142

13 VARIABLE AND UNUSUAL STARS   146

    Exercises: Observed and Physical Characteristics   146
                  Cepheid Variables and the Distance Problem   147
                  Planetary Nebulae   147
                  Exploding Stars   148
    Hints   149
    Solutions   151

14 STELLAR EVOLUTION: THE NIGHT LIFE OF THE STARS   154

    Exercises: Star Formation   154
                  Stellar Energy Production   155
                  Evolution and the H-R Diagram   156
                  Stellar Fates   157
    Hints   159
    Solutions   161

15 THE MILKY WAY GALAXY   166
    Exercises: Shape, Structure and Motions   166
                  Nebulae and the Interstellar Medium   167
                  Relationships of Stars and Galactic Features   169
                  Extra-Terrestrial Intelligent Life in the Galaxy   170
    Hints   171
    Solutions   173

16 GALAXIES AND QUASARS   179

    Exercises: Appearances and Types   179
                  The Local Group of Galaxies   181
                  Galactic Distances   182
                  Energy Output of Galaxies   183
                  Groups and Clusters of Galaxies   183
                  Quasi-Stellar Objects (Quasars)   184
    Hints   185
    Solutions   187

17 COSMOLOGY - THE EVOLUTION OF THE UNIVERSE   193

    Exercises: General Exercises   193
                  The Expanding Universe   194
                  Cosmological Models   195
    Hints   196
    Solutions   197

1 ANCIENT ASTRONOMY

Early Astronomical Concepts

1-1   Cite some purposes for which astronomical knowledge was used by early cultures.

1-2   Cite some reasons why early astronomers rejected the concept of a moving (rotating, revolving) Earth.

1-3   To early astronomers, what was the relationship between astrology and astronomy?

1-4   Why is the concept of the hypothetical celestial sphere of value to modern astronomy?

Astronomical Observations and Phenomena

1-5   Cite some astronomical phenomena that were probably known to ancient man, who lived tens of thousands of years ago.

1-6   Cite some astronomical phenomena observed or predicted by early Chinese astronomers.

1-7   From where on Earth, during the course of one year, can one see

(a) Only half of the celestial sphere,

(b) The entire celestial sphere?

1-8   From where on the Earth do the stars appear to

(a) Move parallel to the horizon,

(b) Rise perpendicular to the eastern horizon and set perpendicular to the western horizon?

(c) On the surface of the Earth, what would be the minimum distance one would have to travel to get from one of your answers to part (a) to your answer to part (b)?

1-9   What simple observation can be made to determine the approximate latitude
       of a place in the northern hemisphere of the Earth?

1-10   How could you determine, with one night's visual observations, that you
       were located at the Earth's north pole?

1-11   List several reasons why people who lived before Christ believed that the
       Earth was round.

1-12   Explain how one could determine the length of the year (as did the ancients)
       from observations of

       (a) The positions on the horizon where the Sun rises and sets each day,

       (b) The length of the shadow cast by a vertical stick in the ground?

1-13   To the early Greeks, what was the significance of the

       (a) Ecliptic,

       (b) Zodiac,

       (c) Constellations?

1-14

Exercise 1-14. (E.C. Krupp, Griffith Observatory)

The photo is of one of the oldest known astronomical observatories.

(a) What is the name of the observatory, and where is it located?

(b) What astronomical phenomena were probably observed or predicted there?

1-15   List several ways in which the planets appear different than the stars in
       the sky.

Important Early Astronomers

1-16   What were some of the astronomical observations and ideas of Aristarchus of Samos?

1-17   What were some of the astronomical contributions made by the Greek Hipparchus?

1-18   (a)   What basic concept did Eratosthenes use when he determined the size of the Earth?

      (b)   What did he observe or measure?

1-19   If Eratosthenes had measured an angle of 14° between the Sun and the zenith at Alexandria, instead of 7°, how would his value for the Earth's size have differed from the value he actually determined?

1-20   Why is Ptolemy's Almagest, written about 1800 years ago, still so important to astronomy today?

HINTS TO EXERCISES ON ANCIENT ASTRONOMY

1-1   The purposes include the prediction of rare astronomical events and/or regularly occurring events.

1-2   The reasons were based on both science and religion.

1-3   The same persons did both astronomy and astrology.

1-4   The celestial sphere is the geocentric sphere of infinite radius upon which we imagine the celestial objects to lie.

1-5
1-6   They are phenomena that involve the objects that can be observed with the naked eye, both then and how.

1-7
1-8   The two answers for part (a) are points on the Earth, and the answer to part (b) is a great circle on the Earth.

1-9   It involves the star that serves as the north star.

1-10   Consider the unique motions and appearances of the stars as viewed from that location. (See Exercise 1-8).

1-11   One reason involves changes in the altitudes of the stars as an observer changes location on the Earth. Another involves the shape of the shadow of the Earth during a lunar eclipse.

1-12   Both phenomena change each day, with a period of one year.

1-13   The first two are related to the path of the Sun as observed against the background stars.

1-14   It is located in England and constructed of large stones.

1-15   They differ in brightness, location in the sky, and relative motions.

1-16   They involve the Sun and the Moon.

1-17   They involve the Sun, Moon, planetary motions, and the stars.

1-18
1-19

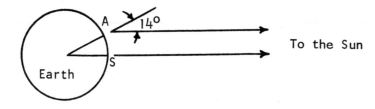

1-20   It is an important historical record.

1-1   Early astronomical knowledge was used for predicting eclipses, construct-
        ing calendars, astrology, religion, and determining seasonal cycles, time,
        location and direction.

1-2   Aristotle could not detect stellar parallax. The Bible says that
        Joshua ordained the Sun to stand still, not the Earth, which implies that
        the Sun moves around the Earth. Authorities and the Church forbad belief
        in a moving Earth. People reasoned that things would fly off a moving
        Earth.

1-3   Many early astronomers were employed to cast horoscopes for the royalty.
        These astronomers made observations for the purposes of astrology, but at
        the same time used the observations for scientific study.

1-4   It allows us to accurately and simply represent the relative locations of
        the celestial objects as we see them in the sky. We represent their direc-
        tions on the celestial sphere but not their distances.

1-5   They were no doubt familiar with the appearances and daily motions of the
        Sun, Moon and stars. They may have known about the changes in the appear-
        ances of these objects: the phases of the Moon, and the seasons due to
        the Sun's yearly motion in the sky. Also, they may have observed comets
        and meteors.

1-6   They predicted lunar and solar eclipses, recorded supernovae, recorded
        observations of Halley's comet, and determined the length of the year.

1-7   (a) At the north or south pole.

        (b) Anywhere on the equator.

1-8   (a) At the north or south pole.

        (b) Anywhere on the equator.

        (c) One-fourth of the Earth's circumference, about 10,000 km.

1-9   The altitude of the north pole star is equal to the latitude of the observer,
        approximately.

1-10   The observation that the stars were moving parallel to the horizon would
        tell you that you were at one of the poles (north or south). Identification
        of a northern hemisphere constellation (a dipper, for example) would place
        you at the north pole.

1-11   Local measurements of the Earth's curvature had been made by Eratosthenes.
        People observed that the Earth's shadow on the Moon (during an eclipse) is
        always curved. The Greeks and Egyptians knew that some of the stars appear
        to rise in the sky as one travels northward on the Earth's surface.

1-12 (a) The rise and set points of the Sun on the horizon change each day, being farthest north in the summer, due east and west at the equinoxes, and farthest south in the winter. Thus the time required for the rise (or set) point to make one full cycle along the eastern (or western) horizon is one year.

(b) In the northern hemisphere draw a straight line on the ground due north from the stick. Each day mark off the point on the line where the tip of the noon shadow crosses the line. This point will go through a yearly cycle--closest to the stick in the summer, farthest away in the winter.

1-13 (a) The ecliptic was the path of the Sun through the heavens.

(b) The zodiac was the band of stars (about 16° wide) through which the Sun moves each year.

(c) The ancient constellations are prominent star groups that were probably named after people, animals, or objects which were important to the cultures.

1-14 (a) Stonehenge in England was constructed during the third millennium BC.

(b) It was probably used to predict lunar and solar phenomena such as seasons, eclipses, and points of rising and setting.

1-15 Planets are usually brighter than most stars and planets twinkle less. Also the planets change their brightness over the course of several weeks; whereas most stars maintain a steady brightness. Over a period of several nights the planets move relative to the stars, and the planets are found only near the ecliptic in the sky.

1-16 He made rough estimates of the distances and sizes of the Sun and Moon, in terms of the Earth's size. He believed that the Sun was at the center of the universe, not the Earth, and he believed that the stars were at great distances.

1-17 He catalogued star positions and established the first system of stellar magnitudes. He charted the Sun's daily positions on the ecliptic and accurately determined the length of the seasons. Hipparchus discovered the precession of the Earth's axis. He determined the size and distance of the Moon, with some accuracy, and he developed eccentrics and epicycles to represent planetary motions.

1-18 (a) He assumed that the Sun was so far away that all the sunlight that strikes the Earth at a given instant has come from the same direction-- the light rays were parallel at all places on the Earth.

(b) He had to measure the distance from Alexandria to Syene, and he measured the angle between the zenith and the noon Sun.

1-19 He would have concluded that the Earth was half as large.

1-20 The Almagest is our primary source of information about early Greek astronomy, since most other records have been lost or destroyed.

Observations and Early Theories

2-1   What are the primary observed features of the motions of the planets in the sky?

2-2   Why do the planets appear (to an Earth observer) in different parts of the sky at different times of the year?

2-3   In what part(s) of the sky, and when, is the planet Venus usually observed?

2-4   What is the historical significance of the word planet?

2-5   Why is the concept of the celestial sphere (upon which all the stars are assumed to lie) useful, even though it is not reality.

2-6   In what way was the universe of Hipparchus and Ptolemy an improvement over the universe of Eudoxus?

2-7   In the universe of Ptolemy,

(a) Where was the Earth located, and how did it move?

(b) Where was the Sun located, and how did it move?

(c) Where were the stars located, and how did they move?

2-8   In the universe of Ptolemy, what was the phase cycle (as observed from Earth) of the planets Mercury and Venus?

2-9   In the universe of Ptolemy, how is retrograde planetary motion explained?

Copernicus, Tycho and Galileo

2-10 (a) What were Copernicus' two major contributions to our present theories of planetary motions?

(b) What scientific evidence did he have to support his ideas?

2-11 The Danish astronomer Tycho Brahe observed a supernova in 1572-73.

(a) How did he determine that the supernova was much farther from the Earth than was the Moon?

(b) Why was this conclusion very important?

2-12 How did each of the following contribute to our understanding of planetary motions?

(a) Tycho's observations of the positions of Mars.

(b) Newton's idea that gravity is universal--that it is a force which acts between any two bodies in the universe, not just on the Earth's surface.

2-13 Which observations provided the first scientific proof that the solar system is heliocentric?

2-14 How did the following observations of Galileo contribute to our understanding of planetary motions?

(a) The satellite system of Jupiter,

(b) The phases of Venus.

2-15 (a) Under what conditions was Galileo forced to recant ideas which he believed were true?

(b) Cite an example of a modern scientific idea that is also politically unfashionable, and therefore subject to ridicule and attack.

Kepler: Motions in the Ellipse

2-16 (a) What were Kepler's contributions to the theory of planetary motions?

(b) Upon what did Kepler base his ideas about planetary motions?

(c) What scientific proof did Kepler have for his ideas?

2-17 What is the semi-major axis of a circle?

2-18 Give two approximate definitions of the astronomical unit.

2-19 Where in an elliptic orbit does a planet move

(a) fastest,

(b) slowest?

2-20 What is the period of a minor planet with a semi-major axis of 4 au?

*2-21 Calculate the eccentricity of the Earth's orbit.

*2-22 It has been discovered that the flying saucers come from the small, dark planet Friedman, which has

aphelion distance = $3\frac{1}{2}$ au,   perihelion distance = $\frac{1}{2}$ au.

(a) Sketch the orbits of the Earth and Friedman around the Sun.

2-22 (b) What is the semi-major axis (in au's) of Friedman's orbit?

(c) What is the period (in years) of Friedman?

*2-23 Using at least half a sheet of paper, and assuming that Venus is in a circular orbit 0.72 au from the Sun,

(a) Sketch the orbits of the Earth and Venus about the Sun, and draw an elliptical trajectory from the Earth to the orbit of Venus which is tangent to both the planetary orbits.

(b) Show the approximate location of Venus on the date of a spacecraft's departure from Earth. Remember that Venus moves faster than the Earth and faster than the spacecraft's average speed.

(c) Calculate the semi-major axis of the spacecraft's trajectory in au's.

(d) Calculate the Earth-Venus trip time in years.

Newton: The Laws of Motion and Gravity

2-24 Cite one astronomical and one non-astronomical example of Newton's first law of motion.

2-25 Cite one astronomical and one non-astronomical example of Newton's second law of motion.

2-26 Cite one astronomical and one non-astronomical example of Newton's third law of motion.

2-27 Each of the following is an example of which of Newton's laws of motion in an elliptic path.
(a) The mutual gravitational forces of the Earth on the Moon and the Moon on the Earth.

(b) A body moving in a straight line at constant speed.

(c) The gravitational force of the Sun causing the Earth to move in an elliptical path.

(d) A rocket expelling gases in a vacuum.

(e) A stone accelerating as it falls to Earth.

2-28 What was Newton's modification of Kepler's third law?

2-29 The gravitational attraction between two bodies changes by how much when the distance between the two bodies is

(a) doubled,

(b) halved,

(c) six times greater?

2-30 How is the gravitational attraction of a planet related to its rotation speed?

2-31 Suppose the gravitational attractions between the Sun and the planets were suddenly turned off.

(a) What would be the subsequent motion of the Earth?

(b) About how long would it take the average man on the street to notice that anything was different, and what would he notice?

2-32 On the Moon's surface the gravitational attraction is about 1/6 of the attraction on the Earth's surface.

(a) What would be the weight on the Moon of a woman who weighed 120 lbs. on Earth?

(b) Describe how the woman's mass would change as she moved from the Earth, through space, to a standing position on the Moon.

HINTS TO EXERCISES ON PLANETARY MOTIONS: HISTORICAL AND MODERN

2-1   The answer involves the motions and brightnesses of the planets.

2-2   All the planets (including Earth) revolve around the Sun at different speeds.

2-3   Planets are best observed against a dark sky. Venus is always observed in the same general region of the sky as the Sun.

2-4   The planets appear to wander among the stars.

2-5   For many problems the stars are used as background reference points.

2-6   Eudoxus proposed that the planets moved on geocentric crystaline spheres.

2-7   Consult a drawing of Ptolemy's geocentric universe.

2-8   The planets are illuminated by the Sun and observed from the Earth.

2-9   The planets move on epicycles which in turn move on large deferents.

2-10   Both ideas dealt with motions of the Earth, and both led to simplified models of the universe.

2-11   He observed the parallaxes of the supernova and the Moon.

2-12   (a) Kepler worked for Tycho, and he used Tycho's observations.

        (b) According to Newton, there is gravitational attraction between the Sun and the planets.

2-13   It was one made by Galileo with his telescope.

2-14   Neither observation could be explained by Ptolemy's model.

2-15   (a) He was forced to give up the idea of a heliocentric solar system.

2-16   Kepler worked with Tycho's observations of the positions of Mars.

2-17   The semi-major axis is half the length of the longest straight line that can be drawn inside an ellipse or circle.

2-18   One involves the size of the Earth's orbit; the other the distance from the Earth to the Sun.

2-19   Use Kepler's second law.

2-20   Use Kepler's third law: $P^2 = a^3$, where P is the period in years, and a is the semi-major axis in au's.

2-21   The Earth-Sun distance varies between 147,100,000 km and 152,100,000 km.

2-22 (c) See hint for Exercise 2-20.

2-23 (d) See hint to Exercise 2-20, and the trip time is half the period.

2-24 The first law describes the motion which occurs in the absence of all unbalanced forces.

2-25 The second law describes the motion that results from the application of a force F to a mass m: $F = ma$, where a is the acceleration of the mass.

2-26 The third law states that forces act in pairs; for every action, there is an equal and opposite reaction.

2-27 Review the basic concept of each of Newton's laws.

2-28 It takes into account the masses of the orbited and orbiting body.

2-29 The gravitational force is $F = G \dfrac{m_1 m_2}{r^2}$

where m_1 and m_2 are the masses of the two bodies which are separated by a distance r.

2-30 See Newton's law of gravity (Hint 2-29).

2-31 Consider Newton's first law, the size of the Earth's orbit, and the speed of the Earth.

2-32 Weight is the force of gravitational attraction on a body; whereas mass is a measure of the amount of matter in a body.

SOLUTIONS TO EXERCISES ON PLANETARY MOTIONS: HISTORICAL AND MODERN

2-1   Primarily the planets move slowly from west to east (direct) relative to the background stars, but once each synodic period they move east to west (retrograde) for a few weeks. They always move near the ecliptic, never in any other part of the sky.

2-2   The planets and the Earth are all revolving around the Sun at different speeds and in different locations. Thus all these bodies are moving relative to each other and to the distant background stars.

2-3   Venus is best observed either in the southwest sky just after sunset, or in the southeast sky just before sunrise.

2-4   It derives from the Greek word for wanderer, since the planets appear to wander among the stars.

2-5   On the celestial sphere the relative directions of objects may be accurately represented without knowing their distances. This is useful because many problems may be solved and many concepts illustrated without knowledge of the distances to objects--the constellations, for example.

2-6   Eudoxus' model never explained the variations in the observed brightnesses of the planets, since their distances from the Earth never changed in his model. In Ptolemy's model, however, the variations were explained as due to changes in the distances of the planets from the Earth and Sun.

2-7   (a) The stationary Earth was at the center of the universe.

       (b) The Sun was located between the epicycle of Venus and the deferent of Mars; it moved around the Earth once a day.

       (c) The stars were located on the great celestial sphere beyond the regions of the planets. The sphere was centered on the Earth, and it rotated once each day.

2-8   For both planets the cycle of phases was new – crescent – new – crescent – new in one revolution.

2-9   The planets show retrograde motion when they are closest to the Earth on their epicycles.

2-10   (a) The rotation of the Earth on its axis, and its revolution about the Sun.

       (b) He had no evidence; he just believed that his scheme was much simpler than any previously proposed.

2-11   (a) He could observe no parallax for the supernova; whereas the Moon exhibits parallax and thus must be much closer.

       (b) This was the first indication that objects on the celestial sphere (stars) underwent change, that they are not forever constant as had been previously believed.

2-12   (a)   Tycho's observations provided the data with which Kepler developed
             his three laws of planetary motion.

       (b)   Newton's law explained <u>why</u> the planets move according to Kepler's
             laws.

2-13   Galileo's observations of the phases of Venus, which could only be explained
       by a heliocentric solar system.

2-14   (a)   In the satellite system he saw a Copernican solar system in the minia-
             ture--a large central body with small bodies orbiting it.

       (b)   The Ptolemaic system predicted that Venus should show only new and
             crescent phases (Exercise 2-8); whereas Galileo observed that Venus
             goes through all the phases, like the Moon and like the Copernican
             theory predicted. This observation proved the error in the Ptolemaic
             theory.

2-15   (a)   He was forced to recant the heliocentric theory of Copernicus and
             Kepler under threat of death or torture by the Roman Inquisition.

       (b)   Darwin's ideas regarding the evolution of man.

             Velikovsky's ideas that collisions have played an important role in
             the evolution of the solar system.

             Shockley's ideas regarding the genetic inferiority of some races.

2-16   (a, b) Kepler developed three laws of planetary motion in the ellipse,
               which were based upon Tycho's observations of Mars.

       (c)   He had no proof that his system was better than Ptolemy's.

2-17   The radius of the circle.

2-18   1)   It is the average distance between the Earth and the Sun.
       2)   It is about 150,000,000 km. or 93,000,000 mi.

2-19   (a)   At perihelion,

       (b)   at aphelion.

2-20 $P^2 = a^3 = 4^3 = 64$, $P = \sqrt{a^3} = \sqrt{64} = $ 8 years.

2-21 $r_p + r_a$ = 299,200,000 km = 2a

       a = 149,600,000 km.

       $r_p = a(1 - e)$

       $\therefore e = \dfrac{a - r_p}{a} = 0.0167$

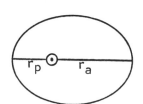

2-22   (a)

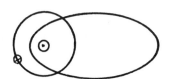

2-22 (b) $2a = \frac{1}{2}au + 3\frac{1}{2} au = 4$ au.

a = 2au

(c) $P^2 = a^3 = 2^3 = 8$, $P = \sqrt{8} = 2\sqrt{2} = 2.83$ years.

2-23 (a)
(b)

Arrival→

0.72 au

Earth at departure

Venus at departure

1.0 au

(c) 2a = 0.72 + 1.0 = 1.72 au.
     a = 0.86 au.

(d) $P^2 = a^3 = (0.86)^3 = 0.636$
     $P = \sqrt{0.636} = 0.80$ years.

Trip time = P/2 = 0.4 years.

2-24 In principle, there are no examples, astronomical or non-astronomical, of Newton's first law since the forces of gravity act throughout the universe and we do not know where they might be in exact balance.

2-25 Astron: The force of the Earth's gravity causes the Moon to move in a curved path around the Earth.

Non-astron: A pitcher's fingers push on the baseball, causing it to accelerate toward the catcher.

2-26 Astron: The Sun exerts a gravitational force on the Earth and the Earth exerts an equal and opposite (in direction) force on the Sun.

Non-astron: A 1 lb book lying on a table exerts a force of 1 lb downward on the table and the table exerts an upward force of 1 lb on the book.

2-27 (a) Third,

(b) First,

(c) Second,

(d) Third,

(e) Second.

2-28 He added the influences of the orbited and orbiting bodies m_1 and m_2 to Kepler's law: $(m_1 + m_2)P^2 = a^3$.

2-29 (a) 4 times weaker,

(b) 4 times stronger,

(c) 36 times weaker.

2-30 They are not related at all. It is a common misconception that if the Earth stopped rotating everyone would fall off. The surface gravity on a planet depends upon only its size and mass.

2-31 (a) The Earth would move in a straight line tangent to its former orbit. Its speed would be constant, equal to its speed at the instant that gravity was turned off.

(b) It would probably take at least a month before he noticed that the Sun was a little smaller and cooler, and that the clocks were not keeping correct time.

2-32 (a) $120/6 = 20$ lbs.

(b) Her mass would not change.

The Bohr Atom

3-1   For each of the following, give the number of protons, electrons and neutrons in one atom (use the most abundant isotope).

(a) hydrogen
(b) deuterium
(c) helium
(d) carbon 12
(e) singly ionized calcium
(f) completely ionized oxygen.

3-2   What are two important roles of the protons in the Bohr atom?

3-3   What are two important roles of the electrons in the Bohr atom?

3-4   Explain what is meant by a(an)

(a) isotope
(b) excited atom
(c) neutral atom
(d) molecule
(e) diatomic molecule
(f) ionized atom
(g) unstable element
(h) radioactive element

3-5   (a) How is light absorbed or destroyed by the atom?

(b) How is this process useful to the astronomer?

3-6   (a) Make a sketch to illustrate how an atom could absorb a photon of energy E_1 and then emit three photons of energies E_2, E_3, and E_4.

(b) What is the mathematical relation between E_1, E_2, E_3, and E_4?

Reflection, Refraction, Dispersion and the Inverse Square Law

3-7   State how light is reflected, and give two examples of reflection in nature: one astronomical, and one non-astronomical.

3-8   State how light is refracted, and give two examples of refraction in nature: one astronomical and one non-astronomical.

Exercise 3-9. Setting Sun. (Photo by Stephen C. Reed.)

What property of light causes the setting Sun (photo) to appear oblate? Make a sketch to illustrate your answer.

3-10 (a) What is dispersion of light?

(b) Give two examples of dispersion of light: one that occurs in nature and one that is man made.

3-11 (a) What happens when white light is passed through a prism?

(b) What happens when light of a single color (monochromatic light) is passed through a prism?

3-12 How many times brighter or fainter does an object (a star, for example) appear if it is moved to

(a) Half its present distance,

(b) Twice its present distance,

(c) Five times its present distance?

3-13 A screen is placed one foot away from a point light source. The screen has a hole in it of area one square inch. Another screen is four feet away from the first screen, on the side away from the light. On the second screen, how large will be the spot of light?

Wavelength, Frequency, Speed and Energy

3-14 State whether the following forms of electromagnetic radiation

radio, TV, infrared, visible, ultraviolet, X, gamma

are listed in order of

(a) increasing or decreasing wavelength,

(b) increasing or decreasing frequency,

(c) increasing or decreasing energy.

3-15 For each of the following forms of electromagnetic radiation, state one way that it is produced in nature, and one way that you might detect the presence of that form of radiation.

   radio,   infrared,   visible,   ultraviolet,   x-rays.

3-16 What is known about the speed of different forms of electromagnetic radiation in a vacuum?

3-17 Refering to the seven kinds of electromagnetic radiation listed in Exercise 3-14, which kind(s)

   (a) is of the highest energy,
   (b) is of the lowest energy,
   (c) pass through the Earth's atmospheric layer,
   (d) is heard by the human ear,
   (e) is received by a car radio,
   (f) causes suntan, sunburn and skin cancer?

3-18 What are the approximate wavelengths of

   (a) red light,
   (b) yellow light,
   (c) blue light,
   (d) radio waves,
   (e) television waves,
   (f) x-rays.

3-19 What is the wavelength of the signals put out by radio station KHJ "Boss Radio", at frequency 930 kilocycles?

3-20 Suppose the distance between ocean waves at Redondo Beach is about 147 feet, and suppose the waves are traveling about ten miles per hour. How often do the waves strike the shore?

3-21 What are two characteristics of sound waves that prove they are not forms of electromagnetic radiation?

3-22 How far (in miles and kilometers) is Mars from the Earth when a radar signal requires 12 minutes 28 seconds for the round trip?

Spectra

3-23 Describe the appearances of the three basic kinds of spectra: emission, absorption and continuum.

3-24 (a) State how each of the three spectra--emission, absorption, continuum-- is produced.

   (b) Cite a specific example of each.

3-25 A glowing cloud of hot hydrogen gas is observed behind a cloud of cool nitrogen gas. Describe and explain the appearance of the spectrum.

3-26 Suppose the spectral energy distribution of a star is observed to have its maximum intensity at the wavelength of 8691 Å. What is the approximate surface temperature of the star?

The Doppler Effect

3-27 Suppose a source is monochromatic yellow light. How does the appearance of the source change as the source

(a) approaches the observer,

(b) recedes from the observer.

3-28 Suppose the observed wavelength of the sodium line in a star is 5891 Å, whereas in the laboratory the wavelength is 5890 Å.

(a) At what speed is the star moving relative to us?

(b) Is the star approaching or receding?

3-29 Describe in words how the Sun's rotation rate can be determined from analysis of the spectra of the light coming from various parts of the solar disk.

3-30 With a statement and sketch, show how the Earth's orbital speed can be determined from observations of the spectrum of a star.

HINTS TO EXERCISES ON ATOMS AND LIGHT

3-1   The number of protons = the atomic number of the atom.
         The number of neutrons = the atomic mass minus the number of protons.
         The number of electrons = the number of protons if the charge of the
                                    atom is neutral.

3-2   The protons are the positively charged, relatively heavy particles in the
         nucleus of the atom.

3-3   Electrons are the small, negatively charged particles that orbit the
         nucleus. Their energy depends upon which orbit they are in.

3-4   The answers involve the following:

                Element that spontaneously emits electrons, gamma rays or helium
                  nuclei.
                An atom that has more protons than electrons.
                A combination of two or more atoms.
                Electrons are at energy levels higher than the lowest possible.
                An atom with additional neutrons in the nucleus.
                An atom with equal numbers of protons and electrons.
                An element which exists as a pair of atoms.
                An element that will spontaneously change into another element.

3-5   Photons of light interact with the electrons in the atom, and the inter-
         actions depend upon the energies of the electrons and photons.

3-6   Use the sketch of the Bohr atom, with the electron in the lowest level
         before it absorbs the photon of energy E_1.

3-7   Both require an additional body or medium, besides just the light, and
3-8   both change the direction of the light rays.

3-9   The rays of light from the setting Sun pass from outer space into the
         atmosphere, and they strike the upper layer of the atmosphere at a sharp
         angle.

3-10   Dispersion is caused by differential refraction--different wavelengths
         are refracted (bent) by different amounts.

3-11   The amount of refraction (bending) depends upon the wavelength of the
         light, the shorter the wavelength the greater the refraction. This
         answer should be sketched.

3-12   Use the inverse square law: $\dfrac{b_1}{b_2} = \dfrac{d_2^2}{d_1^2}$

where b_1 is the brightness at distance d_1, and b_2 is the brightness at
distance d_2.

3-13   The inverse square law applies here too in a slightly different form:

$$\frac{A_1}{A_2} = \frac{d_1^2}{d_2^2} ,$$

where the A's are the areas of the spots of light and the d's are the distances of the screens from the light source.

3-14   Wavelength x frequency = constant (the speed).

Energy is proportional to frequency; i.e., the higher the frequency, the higher the energy.

3-15   All are produced naturally by bodies in the solar system.

3-16   All are the same fundamental phenomena as light.

3-17   The answers are from the following:

Radio,
Visible,
None,
Gamma rays,
Ultraviolet,
Visible and a little ultraviolet.

3-18   The answers are from the following:

10 meters to 10 kilometers,
a few centimeters,
6000 $\overset{\circ}{A}$
5000 $\overset{\circ}{A}$
4000 $\overset{\circ}{A}$
1 to 100 $\overset{\circ}{A}$

3-19   Use the relationship that

Wavelength x frequency = speed of light = 186,000 mi/sec.

Also, 930 kilocycles = 930,000 cycles/sec.

3-20   Use the relationship that wavelength x frequency = speed, and convert the speed to feet per second.

3-21   One characteristic has to do with its speed of travel, and the other with its propagation in a vacuum.

3-22   The radar signal travels at the speed of light, 186,000 mi/sec = 300,000 km/sec.

3-23   These are illustrated in color in most introductory astronomy texts.

3-24   The answers are from among:

A glowing liquid or solid,
A glowing gas,
A cool gas in front of a glowing liquid or solid.

3-25 The hot hydrogen will emit only at wavelengths that are characteristic of hydrogen, and the cool nitrogen will absorb only those wavelengths that are characteristic of nitrogen.

3-26 Use Wien's law: $\lambda_{max}(cm) = \frac{0.2897}{T\ (^{\circ}K)}$,

where T is the surface temperature of the star.

3-27 The appearances are changed by the Doppler effect.

3-28 $\frac{\Delta\lambda}{\lambda} = \frac{v}{c}$, where

λ is the wavelength at rest,
$\Delta\lambda$ is the change in the wavelength due to relative motion,
v is the speed of the object, approaching or receding,
c is the speed of light, 300,000 km/sec.

3-29 Due to the solar rotation, one limb of the Sun approaches the Earth while the other limb recedes.

3-30 If the star is in or near the plane of the Earth's orbit, then the wavelength of the lines in the star's spectrum will change during the year as the Earth periodically approaches and recedes from the star.

SOLUTIONS TO EXERCISES ON ATOMS AND LIGHT

3-1

	Number of Protons	Number of Neutrons	Number of Electrons
a	1	0	1
b	1	1	1
c	2	2	2
d	6	6	6
e	20	20	19
f	8	8	0

3-2   (1)   The number of protons determines the element.

       (2)   The positively charged protons attract the negatively charged electrons and hold them in their orbits.

3-3   (1)   Electrons form the bonds between atoms to give strength to solids and viscosity to liquids.

       (2)   Electrons absorb and emit photons of energy.

3-4   (a)   The atom's nucleus has an abnormal number of neutrons.

       (b)   The electrons are at higher energy levels than the lowest possible.

       (c)   The atom contains an equal number of electrons and protons.

       (d)   A combination of two or more atoms.

       (e)   An element which exists in Nature only as pairs of atoms, such as hydrogen which usually exists as H_2.

       (f)   The atom has lost one or more electrons, so that it has more protons than electrons.

       (g)   An element which will eventually spontaneously change into another element, one of lower atomic number.

       (h)   An element that spontaneously emits energy and/or massive particles from its nuclei.

3-5   (a)   An electron in an atom may absorb an approaching photon of light, provided the photon's energy is "compatible" with the electron.

       (b)   This absorption process is useful because it is selective--which photons get absorbed depend upon the characteristics of the atom. Thus the identity of the absorbing substance can be determined by study of the wavelengths of the light absorbed.

3-6 $E_1 = E_2 + E_3 + E_4$

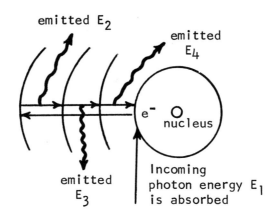

3-7 Reflection is the change of direction of a light ray upon striking a surface without penetration, such that the angle of incidence i is equal to
the angle of reflection r.

Sunlight is reflected off the Moon and
the planets; and it is reflected off snow
or a body of water.

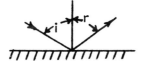

3-8 As light passes from one medium to another, it is bent at the intersections
of the media, such that the light ray is closer to the normal (to the surface of intersection of the media) in the denser medium.

Starlight is refracted as it passes
into the Earth's atmosphere, causing
the stars to appear slightly higher.
Refraction causes a pencil leaning in
a glass of water to appear bent.

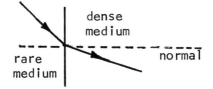

3-9 Refraction. The rays from the lower limb of the Sun are refracted more
than the rays from the upper limb.

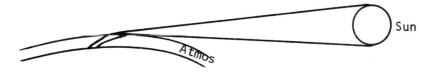

3-10 (a) Dispersion of light is the separation of the light into its various
component colors.

(b) A rainbow is an example of the dispersion of sunlight. The light is
differentially refracted within the raindrops, so that the white
sunlight is spread out into all its component colors. Man disperses
light by passing it through a prism or a grating.

3-11 (a)   white   refraction and
             light   dispersion

(b)   monochromatic   refraction
         light   only

-25-

3-12 (a) $\dfrac{b_1}{b_2} = \dfrac{d_2^2}{d_1^2} = \dfrac{(\frac{1}{2})^2}{(1)^2} = \frac{1}{4}.$ Thus b_2 is 4 times greater than b_1, or the star is 4 times brighter at half the distance.

 (b) 4 times fainter.

 (c) 25 times fainter.

3-13 $\dfrac{A_1}{A_2} = \dfrac{d_1^2}{d_2^2} = \dfrac{1^2}{(1+4)^2} = \dfrac{1}{25}$; $A_2 = 25$ in^2

3-14 (a) Decreasing wavelength (radio is longest).

 (b) Increasing frequency (radio is lowest frequency).

 (c) Increasing energy (radio is lowest energy).

3-15 Radio waves are produced by the Sun and Jupiter in large amounts. They are detected by an ordinary radio or by a radio telescope.

 Infrared is produced by the Sun and by electric stoves and wall heaters. It is sensed by the skin as heat, and it also can be detected by infrared sensitive film.

 Visible light is produced by the Sun, the stars, and any very hot substance. It is detected by film and the human eye.

 Ultraviolet is produced by the Sun, and it is detected by human skin because it causes suntan and sunburn.

 X-rays are produced by exploding stars and the Sun. X-rays are detected by x-ray sensitive film and by devices carried in satellites.

3-16 All forms of electromagnetic radiation move at the same speed in a vacuum, and that speed is about 300,000 km/sec.

3-17 (a) gamma.
 (b) radio.
 (c) visible, radio and some ultraviolet and infrared.
 (d) none
 (e) radio.
 (f) ultraviolet.

3-18 (a) 6000 Å.
 (b) 5000 Å.
 (c) 4000 Å.
 (d) 10 meters to 10 km.
 (e) a few centimeters.
 (f) 1 to 100 Å.

3-19 Wavelength = speed/frequency = $\dfrac{186,000 \text{ mi/sec}}{930,000 \text{ cycles/sec}}$ = 0.20 miles/cycle
 = 1056 feet.

3-20 $10 \text{ mi/hr} = \dfrac{10 \text{ mi} \times 5280 \text{ ft/mi}}{60 \text{ min/hr} \times 60 \text{ sec/min}} = 14.7 \text{ ft/sec}$

Frequency = speed/wavelength

$= \dfrac{14.7 \text{ ft/sec}}{147 \text{ ft/wave}} = 0.1 \text{ waves/sec} = 1 \text{ wave/10 seconds.}$

The waves strike the shore once every 10 seconds.

3-21 (1) The speed of sound in air is about 700 mi/hr; whereas electromagnetic radiation travels at 186,000 mi/sec.

(2) Sound cannot travel through a vacuum (on the Moon, for example); whereas electromagnetic radiation easily travels through a vacuum (light reaching the Earth from the stars).

3-22 The one-way trip takes 6 min. 14 sec. = 374 seconds.

374 sec x 186,000 mi/sec = 69,000,000 miles.

374 sec x 300,000 km/sec = 112,000,000 km.

3-23 Emission: a series of bright, single-colored lines, each of a different color, ordered from red to violet.

Continuum: a continuous unbroken band of light, starting with red, which fades into orange, which fades into yellow, ..., which fades into violet.

Absorption: series of dark lines superimposed on a background of a continuum.

3-24 Emission: a glowing rarefied gas.
Ex: sodium or mercury vapor electric street lights.

Continuum: a glowing solid or liquid.
Ex. an incandescent light bulb.

Absorption: a cool rarefied gas in front of a continuous source.
Ex: the oxygen and nitrogen in the Earth's atmosphere absorb some light from the Sun.

3-25 The observed spectrum will be that of hydrogen in emission. Since nitrogen absorbs only that light that is characteristic of nitrogen, all the light from the hydrogen will pass through the nitrogen cloud unaffected.

3-26 $T = \dfrac{0.2897}{\lambda_{max}} = \dfrac{0.2897}{8691 \times 10^{-8} \text{ cm}} = \dfrac{1}{3} \times 10^{4} = 3333^{\circ}\text{K.}$

3-27 (a) The object's color will be slightly greenish.

(b) The object's color will be slightly orangish.

3-28 $v = c \dfrac{\Delta\lambda}{\lambda} = 186,000 \text{ mi/sec} \times \dfrac{1}{5890} = 31.6 \text{ mi/sec receding.}$

3-29 Since the east limb of the Sun
      approaches the Earth, light
      from it is Doppler shifted toward
      the blue; whereas light from the
      west limb is Doppler shifted
      toward the red. The total dif-
      ference in wavelengths between
      the two limbs ($\Delta\lambda$) will give

$$v = c \frac{\Delta\lambda}{\lambda}$$

where v is twice the surface speed of the Sun at its equator. Finally,
divide the solar circumference by the equatorial surface speed to get
the rotation rate in days/rotation.

3-30 In January the light from the
      star is Doppler shifted toward
      the blue end of the spectrum as
      the Earth moves toward the star;
      whereas in July the star's light
      is shifted toward the red as the
      Earth moves away from it. If $\Delta\lambda$
      is the total difference in wave-
      lengths, then $v = c \frac{\Delta\lambda}{\lambda}$

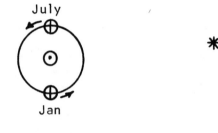

where v is twice the Earth's orbital speed.

Optics and General Characteristics of Astronomical Instruments

4-1   What are the three basic functions of an optical telescope?

4-2   (a)   Define the magnifying power of a telescope.

       (b)   A telescope of 12 inch objective, 84 inch focal length is used with an
             eyepiece of focal length 2 inches. What is the magnifying power of
             this system?

       (c)   Calculate the maximum useful magnifying power of the system.

       (d)   In a given telescope how is the magnifying power increased?

4-3   (a)   What is the light gathering power of a telescope?

       (b)   Calculate the light gathering power of a 16 inch (objective diameter)
             refractor, compared with the human eye with a $\frac{1}{4}$ inch diameter pupil.

4-4   (a)   Define the theoretical resolving power of a telescope.

       (b)   Calculate the theoretical resolving power of a telescope with a 10
             inch objective and a 72 inch focal length.

*4-5   What is the resolving power of the 300 meter radio telescope at Arecibo
       when it operates at the wavelength of 21 centimeters?

*4-6   What is the effective area of the 300 meter radio telescope at Arecibo
       when it observes an object at the zenith?

4-7   (a)   What is meant by the "focal ratio" or "f ratio" of a lens?

       (b)   Calculate the f ratio for the 200 inch Hale telescope at Mt. Palomar.

       (c)   How is f ratio related to the exposure time required to photograph
             an extended object?

4-8   What are the two main functions of the objective lens or mirror of a
       telescope? Illustrate these functions for both the reflector and the
       refractor.

4-9   The objective lens (or mirror) in the refracting (or reflecting) telescope
       is used to bring the light rays from a given point source to the same focal
       point. What two basic laws--one for the refractor and one for the reflector--
       are used to determine the curvatures of the surfaces of the objective lens
       and mirror?

4-10 What are the two main functions of the telescope eyepiece?

4-11 Photographs may be taken or objects may be observed at several different locations on a large reflecting telescope. Identify by name and show by diagram five such locations on reflectors.

4-12 List two comparative advantages and two comparative disadvantages of the reflecting and refracting telescopes?

4-13 (a) What is the principal distortion (aberration) in a refracting telescope? Make a sketch to show why the distortion occurs.

   (b) What are two ways to reduce this distortion?

4-14 Why is chromatic aberration not present in a reflecting telescope?

4-15 (a) What is spherical aberration?

   (b) How may spherical aberration be avoided in a mirror?

4-16 (a) What is the aberration known as coma?

   (b) In what kind of a telescope is it present?

   (c) What is done to reduce the distortion of coma?

4-17

(Lick Observatory Photo)

4-17 (a) What kind of telescope is shown in the photo?

      (b) On the photo, locate the objective and the eyepiece.

4-18

The 200-inch Hale telescope.   (Hale Observatory photo.)

      (a) What kind of telescope is shown in the photo?

      (b) On the photograph locate the objective, the observer's cage, the
           direction to the north celestial pole, and the direction to the coudé
           room.

      (c) Toward what part of the celestial sphere is the telescope pointed as
           shown?

Observing With Astronomical Instruments

4-19 For each of the following, what is the best kind of instrument to use?

      (a) Determine the exact distance to the Moon.
      (b) Determine the rotation rate of Mercury.
      (c) Photograph very faint galaxies.
      (d) Determine the specific wavelength of the light emitted by a galaxy.
      (e) Photograph a large region of the sky, such as the Pleiades or the
           Andromeda galaxy.
      (f) Study the long-wave, low energy radiation emitted by the Sun and Jupiter.
      (g) Resolve the components of a close binary pair.
      (h) Observe the radiation emitted by very cool stars.
      (i) Observe x-rays from a black hole.

4-20 As viewed through a telescope, how do stars appear different than they do as viewed with the naked eye?

4-21 Does one use the eyepiece to do astronomical photography? Explain your answer.

4-22 What are the advantages of photographic astronomical observations as compared with visual observations?

4-23 Why are Schmidt type cameras especially suited for photographing large areas of the sky?

4-24 What are some advantages of a radio telescope, compared with an optical telescope of equal objective size?

4-25 What are some advantages of radar astronomy over optical astronomy?

4-26 List some of the objects in the solar system which have been detected (or observed) by Earth-based radar.

4-27 Compared with ground-based observations, what are the advantages of making astronomical observations from artificial earth satellites?

4-28 How is it possible to make astronomical observations of infrared radiation during the daytime?

4-29 What is the advantage of laser astronomy over optical astronomy?

4-30 If a star is too faint to be observed visually with a telescope of 12 inch objective and 72 inch focal length, and a 1 inch eyepiece, what can be done to observe the star?

4-31 (a) List three ways of gathering astronomical data (making astronomical observations) which do not involve the use of ground-based optical or radio telescopes.

(b) Why is it necessary to use these non-traditional observing methods?

*4-32 If the separation between two binary stars is $0''.228$, what is the smallest telescope (objective diameter) that can theoretically resolve the pair?

*4-33 At the November 1976 opposition, the equatorial angular diameter of Jupiter was $45''.68$. What would be the size of Jupiter's image as photographed with the 200 inch Hale telescope (objective = 200 inches, focal length = 660 inches)?

*4-34 (a) Suppose Jupiter at opposition has an angular diameter of $50''$. If you wished to observe Jupiter as the same size as the Moon (angular diameter = $\frac{1}{2}^\circ$), what magnifying power would be necessary?

(b) If the telescope you are going to use to observe Jupiter in the above situation has a 6 inch objective and a 72 inch focal length, what is the focal length of the eyepiece that is needed?

Auxiliary Astronomical Equipment

4-35   (a)   What is a guide scope?

        (b)   Why do moderate and large telescopes need guide scopes?

4-36   (a)   What is the role of the clock drive on a telescope?

        (b)   Why does an equatorial telescope need only one clock drive (instead of two)?

4-37   (a)   What is the function of the photoelectric photometer?

        (b)   How does it improve upon or supplement visual and photographic observations?

4-38   (a)   What is the function of the spectrograph?

        (b)   How does the spectrograph improve upon or supplement visual and photographic observations?

Locations of Astronomical Instruments

4-39   Why are optical telescopes usually located on remote mountain tops?

4-40   Compared with an Earth-based telescope, what would be two advantages and two disadvantages to having an optical telescope on the Moon?

4-41   Compared with having an astronomical telescope on the Moon, what is one advantage and one disadvantage of having a telescope in an Earth-orbiting satellite?

4-42   Why are radio telescopes usually located in remote valleys?

4-43   Why is the Cal Tech Solar Observatory located in Big Bear Lake?

44-4   Why is the Kuiper Observatory located in a jet aircraft?

4-1   Another way to ask the question is: what are the three changes that the telescope has to make in the appearance of a distant object in order to make it appear closer?

4-2   Magnifying power = $\dfrac{\text{focal length of objective}}{\text{focal length of eyepiece}}$.

The maximum useful magnifying power is approximately 50D. D is the objective diameter in inches.

4-3   The area of a circle is $\pi(d/2)^2$, where d is the diameter of the circle.

4-4   Resolving power is expressed as $d = 4.56''/a$ where d is the angular separation of two stars (in arc seconds) and a is the objective diameter in inches.

4-5   For a radio telescope, resolving power is $70°\lambda/D$ where λ is the wavelength of operation and D is the antenna diameter in the same units as λ.

4-6   The area of a circle is πR^2, where R is the radius.

4-7   The 200 inch Hale telescope has a focal length of 660 inches.

4-8   One function is related to the size of the objective, the other to its curvature.

4-9   The lens and mirror bring light to a focus by changing its direction of travel, so the laws are those that describe the two ways that the direction of light may be changed.

4-10   To get at this question, consider how the light rays are changed as they move from the focal plane, through the eyepiece, to the eye.

4-11   The locations are the focal points, where the light is focused to a point.

4-12   Some of the comparisons have to do with size, ease of construction, and distortions.

4-13   The distortion is due to the dispersion of light of different colors as it is refracted through a lens.

4-14   Chromatic aberration is due to the dispersion of light as it is refracted through a lens.

4-15   Spherical aberration is found in mirrors that have spherical surfaces.

4-16   An ideal astronomical photograph should represent all the stars as perfectly spherical points.

4-17   This telescope uses an objective lens.

4-18 The objective is a large mirror.

4-19 The answers include: large reflector, large refractor, x-ray detector above the Earth's atmosphere, radar telescope, radio telescope, infrared telescope, laser telescope, spectroscope, and Schmidt camera.

4-20 The answers to this question are related to the answers to Exercise 4-1, but remember the telescope cannot magnify the images of the stars.

4-21 The photographic plate is located at the focal plane of the objective.

4-22 Both these answers have to do with time.

4-23 The Schmidt camera is a reflecting telescope with a correcting lens to reduce the distortions of coma.

4-24 Consider what kinds of objects are detected by the two instruments.

4-25 In radar astronomy, radio signals are both sent and received by the observer.

4-26 The objects include the inner planets, some asteroids, and some satellites.

4-27 The satellites are in orbit above the Earth's atmosphere.

4-28 Infrared radiation is not visible.

4-29 In laser astronomy, a very narrow light beam is bounced off a distant object.

4-30 There are two solutions; one involves photography with the telescope in the question, the other involves use of another telescope.

4-31 These are ways of observing outside the Earth's atmosphere.

4-32 See hint of Exercise 4-4.

4-33 $A = 0.000004844 \ f \ d$, where

A is the image size,
f is the telescope focal length in the same units as A (usually inches),
d is the angular size of the object (arc seconds).

4-34 See hint for Exercise 4-2.

4-35 Large telescopes usually have very narrow fields of view, and astronomers often photograph objects that cannot be directly observed.

4-36 It is convenient if the telescope on the moving Earth can be kept pointing at the same point in the sky for long periods of time.

4-37 This instrument counts photons of light.

4-38 The spectrograph uses a prism or grating to disperse light.

4-39 The two factors that lead to optimum observing conditions are (1) a dark sky, and (2) a minimum amount of distortion due to the atmosphere.

4-40 There is no atmosphere on the Moon.

4-41 The answers to Exercise 4-40 give some clues to this one.

4-42 Radio telescopes operate best in the absence of man-made radio signals.

4-43 A given amount of solar radiation heats water less (to a lower temperature) than it heats land.

4-44 Many types of electromagnetic radiation do not pass easily through the Earth's atmosphere.

SOLUTIONS TO EXERCISES ON ASTRONOMICAL INSTRUMENTS

4-1 (1) The telescope must gather light to make the object appear brighter.

(2) The telescope must magnify the image; to make an extended object appear larger, or to make several point sources appear farther apart.

(3) The telescope must clarify or resolve; to make extended objects more distinct, to allow more detail to be seen.

4-2 (a) Magnifying power is the increase in the angular size of an object as seen through the telescope, compared with its angular size as seen with the unaided eye. For example, the Moon has an angular diameter of $\frac{1}{2}^\circ$ as seen with the unaided eye. If it has an angular diameter of 4° as seen with a telescope, then that telescope and eyepiece are said to have a magnifying power of 8.

(b) Mag. Power = 84/2 = 42.

(c) Max. magnifying power = 50 D = 50 × 12 = 600 power.

(d) Magnifying power depends upon the focal lengths of the objective and the eyepiece. Since the objectives of most telescopes cannot be changed, and since eyepieces are very easy to change, the change to an eyepiece of shorter focal length is the easiest way to increase a telescope's magnifying power.

4-3 (a) The light gathering power is the ability of the telescope to gather light, compared to some other light-gathering device (the eye, for example). It is numerically equal to the area of the telescope's light gathering lens (objective) divided by the area of the comparison device.

(b) $\text{L.G.P.} = \dfrac{\text{area of objective}}{\text{area of eye pupil}}$

$= \dfrac{\pi\left(\frac{16}{2}\right)^2}{\pi\left(\frac{1}{8}\right)^2} = \dfrac{64}{1/64} = 4096$

Theoretically then, the 16 inch telescope gathers 4096 times more light than one human eye.

4-4 (a) Resolving power is the angular separation of the closest pair of stars that can be seen as two stars (resolved) by the telescope. Any binary pair that is closer together than the resolving power is seen as a single star (unresolved).

(b) Resolving power $d = \dfrac{4\overset{''}{.}56}{a} = \dfrac{4\overset{''}{.}56}{10} = 0\overset{''}{.}456.$

Thus if a binary pair are separated by $0\overset{''}{.}456$ or more, they can be resolved by a 10 inch telescope. The focal length has nothing to do with resolving power.

4-5 Resolving power = $70^\circ\ \dfrac{\lambda}{D} = \dfrac{21\text{ cm}}{300 \times 10^2\text{ cm}} \times 70^\circ = 0.049^\circ.$

4-6   For a radio telescope, the effective area is just the area of the antenna
       as seen by the object under observation. In this case it is the area of
       the dish if it were a flat plate.

       Area = πR^2 = $\pi(150)^2$ = 70686 sq. meters.

4-7   (a) The quantity f/a is called the focal ratio or f ratio, where a is
            the objective diameter and f is the focal length of the objective.
            The f ratio is a measure of a telescope's ability to create a bright
            image of an extended source, since the image's brightness is propor-
            tional to $(a/f)^2$.

       (b) For the 200 inch telescope at the prime focus, the focal ratio is
            a/f = 200/600 = 1/3.3. This ratio is often expressed as f/3.3.

       (c) The smaller the f ratio the "faster" the telescope, or the shorter
            the required exposure time.

4-8   The objective gathers a large amount of light; hence the larger the object-
       ive, the fainter the object that may be observed.

       The objective focuses the gathered light to points on the focal plane.

4-9   The law of refraction is used to determine the curvature of the surfaces
       of the lens.
       The law of reflection is used to determine the curvature of the mirror.

4-10   The eyepiece causes all the light rays from a given point source to enter
       the eye from the same direction, as parallel rays.

       The eyepiece magnifies the image by causing light rays from two points to
       enter the eye from more separate directions.

4-11

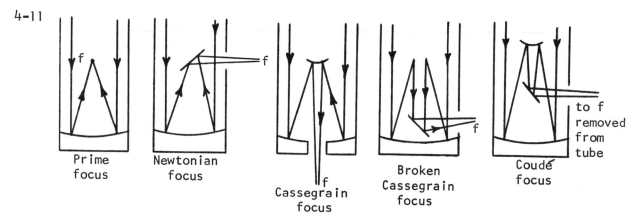

-38-

4-12 Reflector
  Advantage: Can be used at several different focuses.
                 Mirror is easier to build than a lens.
                 Telescope is easier to build in large sizes.
                 More compact, shorter length.

  Disadvantage: Suffers from coma.
                 Transmits less light than a refractor.

  Refractor
  Advantage: Image is more accessible.
                 Larger undistorted field of view.

  Disadvantage: Lens is hard to build.
                 Suffers from chromatic aberration
                 Long length requires large observatory dome.
                 Large lens tends to sag since it is only supported around
                   edges.

4-13 (a) The distortion is chromatic aberration. Short wavelength light is
         focused closer to the lens than light of long wavelength, even though
         the light is from the same point source.

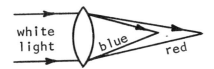

     (b) One way to reduce chromatic aberration is to increase the focal length
         of the refractor. Another way is to use an objective lens that is made
         of several adjacent pieces of different kinds of glass. Each kind of
         glass has different refractive properties, and their combination can
         serve to reduce chromatic aberration. A third way is to use a filter
         to reduce the range of wavelengths that are observed.

4-14 Chromatic aberration is due to refraction of different colors by different
     amounts, and light is not refracted by the mirror in a reflecting telescope.

4-15 (a) Spherical aberration is a defect that is characteristic of mirrors that
         use spherical surfaces (which happen to be the easiest kind to grind).
         It causes parallel beams of light that strike near the edge of the
         mirror surface to be focused at a point closer to the mirror; whereas
         parallel beams that strike the mirror near its center are focused
         farther from the mirror. An ideal mirror should focus all parallel
         rays at the same point.

     (b) Spherical aberration is eliminated by grinding the mirror surface to a
         paraboloid of revolution.

4-16 (a) Coma is caused by the failure of a parabolic mirror to reflect off-axis
         parallel rays (from sources that are not exactly in the center of the
         field of view) to exactly the same point. This causes stars near the
         edges of a photo to appear slightly elongated or even as hooks or commas.

4-16 (b) This distortion is present in reflecting telescopes. Examples of coma can be seen in the corners of some of the well known, beautiful photos from the Hale Observatory: the Lagoon Nebula in Sagittarius and the globular cluster M-13 in Hercules, for example.

(b) Coma is avoided by (1) working only near the center of the field of view, or (2) using a Schmidt camera which is designed specifically to avoid coma.

4-17 (a) The telescope is a long-focus refractor.

(b) The objective is at the upper right, the eyepiece at the lower left.

4-18 (a) It is a reflecting telescope.

(b) The objective is in the lower center of the photo, and the cage is inside the main tube at the upper center of the photo. The north celestial pole is directed from lower left to upper right, along the large support tube in the foreground; the coudé room is in the opposite direction.

(c) Toward the zenith.

4-19 (a) Laser with a large reflector.
(b) Radar antenna.
(c) Large reflecting telescope.
(d) Spectrograph.
(e) Schmidt camera.
(f) Radio telescope.
(g) Large refractor.
(h) Infrared telescope, on a high mountain or in orbit.
(i) X-ray telescope in orbit.

4-20 (1) The naked-eye stars appear brighter as seen through the telescope.

(2) Many stars that are invisible to the naked eye are seen through the telescope.

(3) Some stars that appear single to the naked eye will appear as multiple stars as seen with the telescope.

4-21 No. For photographic observations the eyepiece and its holder are replaced by the plate and the plate holder, located so that the photographic emulsion on the plate is exactly in the focal plane.

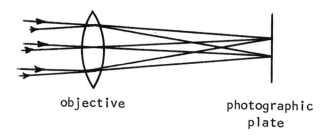

objective   photographic
                                plate

4-22 (1) Photographs can be exposed for a long period of time, thus enabling very faint objects to be recorded. The eye, on the other hand, records things only as they appear at each instant.

4-22 (2) Photos provide a permanent record that can be studied by many astronomers over a long period of time.

4-23 The corrector plate in the Schmidt camera removes the distortions of coma, so the Schmidt camera provides a distortion-free picture of a large area of the sky.

4-24 (1) Radio telescopes examine parts of the spectrum that are not visible to the eye. Some objects in space, neutral Hydrogen for example, give off radio waves but no visible light, so they can best be studied by a radio telescope.

(2) It is possible to operate a radio telescope at all times, not just when it is clear and dark.

4-25 (1) Radar has enabled the astronomer to measure interplanetary distances much more accurately than optical methods.

(2) Radar has enabled the astronomer to map the surfaces of Mars and Venus in some detail, and to determine the rotation rates of Mercury and Venus.

(3) Also, radar bounced off the rings of Saturn has provided an estimate of the size of the ring particles.

4-26 The objects include the Sun, Mercury, Venus, Mars, the Moon, Saturn's rings, the satellite Ganymede, and the asteroids Eros, Icarus and Toro.

4-27 Astronomical observations from satellites are not hindered by the Earth's atmosphere. The atmosphere prevents gamma rays, x-rays, some ultraviolet, some infrared, and some radio waves from reaching the Earth's surface, so these must be studied above the atmosphere.

4-28 Since infrared radiation is not visible, the visible daylight does not interfere with infrared observations, just as light does not interfere with radio observations.

4-29 Laser astronomy can determine short distances (such as from the Earth to the Moon) to much greater accuracy than either radar or traditional optical methods.

4-30 (1) Using the same telescope, a camera could be used to take a time exposure of the star which would bring out fainter objects than are seen by the eye.

(2) On the other hand, if visual observations are required, then another telescope is necessary, one with a larger objective to give more light-gathering power.

4-31 (a) Using instruments on satellites, spacecraft, rockets, balloons, and left on the Moon.

(b) These are needed to observe radiation that does not pass through the atmosphere, such as X-rays. Also, spacecraft are required to observe the planets at close range, and to sample the interplanetary environment.

4-32 Resolving power = minimum separation = $\dfrac{4\overset{''}{.}56}{a}$.

or

$a = \dfrac{4\overset{''}{.}56}{\text{min. separation}} = \dfrac{4.56}{0.228} = 20$ inches.

Thus a 20 inch telescope or larger will resolve this binary pair. Smaller telescopes will see the pair only as a single star.

4-33 Image size (inches) = 0.000004844 x focal length (inches)

x angular diameter of object (arc secs.)

= 0.000004844 x 660 x 45.68

= 0.146 inches.

4-34 (a) If Jupiter's true angular size is 50" and you want it to appear as $\frac{1}{2}°$ (which is 1800"), then Jupiter's image must be increased by 1800/50 = 36. A 36 power telescope system is required.

(b) Mag. power = f_o/f_e
or

$f_e = f_o/$mag. power = 72 inches/36 = 2 inches.
Thus an eyepiece of 2 inch focal length is required.

4-35 (a) A guide scope is a relatively small, wide field telescope that is mounted so as to point in the same direction as the large telescope it accompanies.

(b) Large telescopes have very small fields of view in which it is difficult to locate small objects of interest. If the object is first located and then centered in the guide scope, then it usually is in the field of the large telescope.

For photography, the guide scope is used to view the object and keep the system pointing in the correct direction while the large telescope takes the photo.

4-36 (a) The clock drive moves the telescope in right ascension at the rate of the Earth's rotation (360° per 24 hours), so that the telescope is automatically kept pointing in the same direction in space for a long period of time, to permit extended examination of the object.

(b) An equatorial telescope has one axis aligned parallel to the Earth's rotation axis. This is the polar axis of the telescope. With this arrangement the telescope need be moved only about the one polar axis (with one clock drive) to compensate for the Earth's motion.

4-37 (a) This device electronically measures the amount of light received from a source.

(b) The photometer is an improvement upon visual and photographic measurements because it is much more accurate.

4-38 (a) The spectrograph divides light up into its component colors and photographs the resulting spectrum, thus enabling study of the individual components of the light beam.

4-38 (b) Visual or photographic observations study either all the light components together, or, with the use of filters, they may study individual components. The spectrograph enables study of all the components together.

4-39 (1) To be away from man-made lights. Artificial light reflects off the atmosphere to produce a milky gray sky instead of the ideal black sky.

(2) Optical telescopes are placed at high altitudes to reduce the amount of atmosphere through which the astronomer must observe. The less atmosphere to look through, the fewer atmospheric distortions.

(3) On mountain tops, the observatories have an unobscured horizon, so objects may be viewed to very low altitudes.

4-40 Advantage: No atmosphere to cause distortions.
No city lights to lighten the sky.
Brighter objects can be observed through the lunar day.
The slowly rotating Moon is a very stable observing platform.
The long lunar nights (14 Earth days) allow very long exposures.

Disadvantage: Very expensive to build and maintain.
Facilities would be exposed to meteorite impacts, harmful solar radiation, and large temperature variations.

4-41 The advantage would be that the Earth-orbiting telescope would be much easier to maintain and build.

The disadvantage is that the satellite is not a very stable platform, and periods of continuous darkness are much less than 14 days.

4-42 (1) Radio telescopes are in remote areas to be away from the man-made radio signals that interfere with the signals from outer sapce.

(2) They are placed in valleys because the surrounding hills and mountains serve to block out man-made radio signals.

4-43 During the daytime, when the Sun is observed, the water around the solar observatory is cooler than surrounding land would be. Thus there is less heating and movement of the air around the observatory, and hence less distortion of the observed solar images.

4-44 The Kuiper Observatory is used to observe infrared radiation which is best detected at high altitudes. Also it is used to make observations above remote regions of the Earth.

Properties of the Earth's Interior and Surface

5-1   From what two sources do we obtain most of our information about the interior of the Earth?

5-2   (a) What is presently believed to be the internal structure and composition of the planet Earth?

      (b) For each of the layers, give its approximate size or depth and the two most abundant elements.

5-3   Determine the approximate latitude and longitude (within a degree) of the following locations:

      (a) Royal Greenwich Observatory,
      (b) Hale Observatory, Mt. Palomar,
      (c) North Pole,
      (d) Equator,
      (e) South Pole,
      (f) Your home.

5-4   From where on the Earth's surface can one walk 10 miles due south, 10 miles due west, 10 miles due north, in that order, and be back exactly where one started the trip?

5-5   List several pieces of evidence that indicate (but do not prove) that the Earth has a spherical shape.

5-6   List several proofs that the Earth has a spherical shape.

5-7   List several pieces of evidence that indicate (but do not prove) that the Earth rotates on its axis.

5-8   List several proofs that the Earth rotates.

5-9   (a) What is the definition of a planet's oblateness?

      (b) What is the value of the Earth's oblateness?

      (c) What causes a planet to be oblate?

*5-10   (a) Assuming that the Earth is a perfect sphere of radius 6370 kilometers, calculate the circumference and volume of the Earth, in both kilometers and miles.

      (b) If the average density of the Earth is 5.5 gm/cm^3, calculate the mass (in grams) of the Earth.

5-11 (a) Discuss the advantages or disadvantages of locating a Foucault pendulum at the north pole and on the equator.

(b) How long does it take for the pendulum to complete one revolution at these two locations?

*5-12 By what percent is a person's weight decreased by moving from sea level to the top of a two-mile high mountain? Assume the Earth is a perfect sphere of radius 4000 mi.

Properties of the Earth's Atmosphere and Beyond

5-13 What are the two most abundant elements in the Earth's atmosphere, and what are their relative abundances?

5-14 The universe is believed to be about 75% Hydrogen, yet why is there so little Hydrogen in the Earth's atmosphere?

5-15 Most of the Earth's atmosphere is less than 100 miles above the surface. On a standard 12 inch (diameter) globe of the Earth, how thick would be the 100 mile layer of the atmosphere?

5-16 (a) What causes the aurora?

(b) Why are aurora observed primarily in the high latitude regions of the Earth?

5-17 What is the general size, structure and composition of the Van Allen radiation belts around the Earth?

5-18 (a) How were the Van Allen radiation belts discovered?

(b) How are the Van Allen radiation belts studied?

Influence of the Atmosphere on Astronomy

5-19 (a) What is meant by the refraction of starlight?

(b) At what altitude in the sky is refraction a minimum, and where is it a maximum?

5-20 Why is it difficult to detect Oxygen in the atmospheres of other planets and of stars?

5-21 What causes stars to twinkle?

5-22 Why do planets appear to twinkle less than stars?

5-23 To an observer on the surface of the Earth, how does the atmosphere affect the nighttime appearance of a star

(a) At the zenith,
(b) Near the horizon,
(c) Midway between the horizon and the zenith?

5-24 On a dark, moonless night over 3000 stars may be seen with the naked eye under the best observing conditions. When the gibbous or full moon is in the sky, however, all the fainter stars are invisible to the naked eye. Explain why only the brighter stars are visible on a bright moonlit night.

5-25 Why are only the brightest stars visible in the night skies of the large cities.

5-26 (a) Why are stars and planets not usually visible to the naked eye in the daytime?

(b) Why is it often possible to see Venus in the daytime from a location high in the mountains?

5-27 Describe and explain the changes that occur in the observed color and shape of the Sun as it sets.

5-28 (a) In what ways does the atmosphere hinder the astronomer?

(b) In what ways does the atmosphere help the astronomer?

5-1   One source measures the attraction of the Earth's gravity; the other involves shock waves in the interior.

5-2   The Earth is believed to have four distinct layers.

5-3   (a) is in England,

       (b) is in Southern California.

5-4   The solution set is infinite!

5-5   The evidence falls into two categories:
       (1) Observations of bodies that are similar to the Earth,
       (2) Indications of local curvature.

5-6   The proofs fall into two categories:
       (1) Observations of motions of bodies that travel around the Earth,
       (2) Theoretical studies.

5-7   The evidence lies in objects that we observe in the sky and in information from "authoritative sources."

5-8   A proof of rotation is a phenomenon that can only be explained as due to the Earth's rotation.

5-9   Oblateness is related to the difference between the polar and equatorial diameters of a planet. For the Earth it is small.

5-10   If the radius of a sphere is R, then

       circumference = $2\pi R$,   volume = $\frac{4}{3}\pi R^3$.

       Mass = volume x density, and 1 kilometer = 0.6214 miles.

5-11   At one of the locations the experiment doesn't work; at the other there's no one to observe it.

5-12   The force of gravity between two bodies of masses m_1 and m_2 is given by

$$F = G\,\frac{m_1 m_2}{r^2},$$

       where r is the separation distance between the bodies and G is the gravitational constant. Also, on Earth a person's weight is just the gravitational force of the Earth on the person.

5-13   One element is necessary for life as we know it; the other is an inert gas.

5-14   Consider what are the requirements for a planet to retain any kind of atmosphere.

5-15 On a 12 inch globe of the Earth, 1 inch = 666 miles.

5-16 Aurora are related to solar activity, and are often called "northern lights."

5-17 These are regions that surround the equatorial regions of the Earth, several hundred miles above the surface.

5-18 These cannot be observed from the Earth's surface; they have to be directly sampled.

5-19 Consider the law of refraction as applied to a beam of light passing from outer space into the atmosphere to the observer.

5-20 Elements are detected by analysis of the spectrum of the body in question.

5-21 Most stars themselves shine with a steady light.

5-22 Stars are point sources of light in the sky; whereas planets are extended light sources.

5-23 Consider the effects of refraction and scattering of light from the star.

5-24 Consider atmospheric scattering of the moonlight.

5-25 Consider the effect of the atmospheric reflection and scattering of the light from the city.

5-26 Consider why the sea-level sky is blue and why the mountain sky is even darker blue.

5-27 As the Sun sets its light passes through more and more air to reach the observer, and the rays strike nearly tangent to the upper layer of the atmosphere. See Exercise (3-9).

5-28 Several hindrances have been mentioned in previous questions. The helps are basically those that keep him alive.

SOLUTIONS TO EXERCISES ON THE EARTH - A PHYSICAL BODY

5-1 (1) Detailed information about the Earth's gravity field, and therefore its mass distribution, is obtained from study of the motions of close artificial Earth satellites.

(2) Earthquakes produce shock waves within the Earth, which are studied to learn about the interior.

5-2

	Size	Main Constituents
Crust	3-30 mi thick	Oxygen, Silicon
Mantle	1800 mi thick	Silicon, Iron
Outer liquid core	1000 mi thick	Nickel, Iron
Inner solid core	2000 mi dia.	Nickel, Iron

5-3

	Latitude	Longitude
(a)	$+51°$	$0°$
(b)	$+33°$	$117°$ west
(c)	$+90°$	undefined
(d)	$0°$	$0°$ to $360°$
(e)	$-90°$	undefined
(f)	depends upon where you live	

5-4 (1) The most obvious solution is to begin at the north pole; from there any direction you walk is due south.

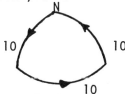

(2) Less obvious solutions are found in the region of the south pole.

5-5 (1) All other planets are observed to be spherical.
(2) Eratosthenes' measurement of the local curvature.
(3) Ships that disappear at sea indicate local curvature.

5-6 (1) The dynamics of artificial Earth satellites.
(2) Observations of the Earth made by the astronauts.
(3) Changes in the locations of constellations in the sky as one travels N-S on the Earth's surface.
(4) The Earth has been circumnavigated in most directions by boats and airplanes.
(5) The Earth's shadow on the Moon during lunar eclipses is always curved.
(6) Theoretical studies of rotating bodies prove that a body with the mass, density and rotation rate of the Earth must be a sphere.

5-7 (1) Every solid body and most gaseous ones in the universe are observed to rotate, including all the other planets and the Sun.
     (2) The daily motions of the Sun and the stars across the sky.
     (3) Books, teachers and parents say the Earth rotates.

5-8 (1) The oblateness of the Earth.
     (2) The Foucault pendulum experiment.
     (3) The directions of ocean and wind currents.
     (4) Observations of the Earth made by the astronauts.
     (5) The observed motions of communications satellites in 24-hour orbits.
     (6) The fact that the Earth precesses.

5-9 (a) Oblateness $= \dfrac{\text{equatorial diameter} - \text{polar diameter}}{\text{equatorial diameter}}$

     (b) The Earth's oblateness $= 1/297 = 0.00337$.

     (c) Oblateness is caused by a planet's rapid rotation.

5-10 (a) Circ. $= 2\pi R = 2\pi \times 6370$ km $= 40,020$ km.

     R(mi) $= 6370$ km $\times 0.6214$ mi/km $= 3958$ mi.

     Circ. $= 2\pi \times 3958$ mi $= 24,870$ mi.

     Vol. $= \dfrac{4}{3}\pi R^3 = \dfrac{4}{3}\pi(6370 \text{ km})^3 = 1.083 \times 10^{12}$ km^3.

     $\qquad = \dfrac{4}{3}\pi(3958 \text{ mi})^3 = 2.597 \times 10^{11}$ mi^3.

     (b) Density $= 5.5$ gm/cm$^3 \times (10^5 \text{ cm/km})^3 = 5.5 \times 10^{15}$ gm/km^3.

     Mass $=$ Volume $\times$ density $= 1.083 \times 10^{12}$ km$^3 \times 5.5 \times 10^{15}$ gm/km^3

     $\qquad\qquad = 5.957 \times 10^{27}$ gm.

5-11 (a) At the north pole the experiment works best but no one would see it. On the equator the experiment does not work.

     (b) 24 hours at the pole: forever at the equator.

5-12 The sea level weight of the person (m_1) is

$$F_{s1} = G\,\frac{m_1 m_e}{(4000)^2}$$

On the two-mile mountain the weight is

$$F_{mt} = G\,\frac{m_1 m_e}{(4002)^2}$$

Where m_e is the mass of the Earth.

Thus the ratio of the two weights is $\dfrac{F_{s1}}{F_{mt}} = \dfrac{\left(\frac{1}{4000}\right)^2}{\left(\frac{1}{4002}\right)^2} = \left(\dfrac{4002}{4000}\right)^2 = \left(1 + \dfrac{2}{4000}\right)^2$

$\approx 1 + \dfrac{1}{1000}$

(continued on next page)

5-12   The sea level weight is 1/1000 greater than the weight on the mountain, or the mountain weight is 0.1% less.

5-13   The atmosphere is composed mainly of Nitrogen (78%) and Oxygen (21%). Many gases make up the remaining one percent.

5-14   Hydrogen is the lightest known gas, and the Earth has insufficient surface gravity to keep the light Hydrogen molecules from escaping into space.

       To put it another way, at the temperatures of the Earth's atmosphere the Hydrogen molecules have enough energy to overcome the surface gravity and escape into interplanetary space.

5-15   If 12 inches represents the Earth's 8000 mile (approximate) diameter, then one mile is 12/8000 inches, and 100 miles is 100 x (12/8000 inches) = 0.15 inches.

5-16   (a) Aurora is the light produced when charged particles or high-energy radiation from the Sun interact with the Earth's upper atmosphere.

       (b) Aurora occur in the polar regions because the charged particles follow the lines of force in the Earth's magnetic field, and those lines dip into the atmosphere in the polar regions.

5-17   The two belts are concentric, donut-shaped regions that surround the Earth and lie near the plane of the magnetic equator. The regions are concentrations of charged particles--high energy electrons and protons.

       The inner belt extends from about 500 miles 3500 miles above the Earth's surface, and the outer belt from 10,000 to 15,000 miles altitude.

5-18   (a) The radiation belts were discovered by the first U.S. satellites, carrying high-energy radiation detection devices that were designed by James Van Allen.

       (b) The radiation cannot be observed from the ground, so the belts must be studied by artificial satellites.

5-19   (a) Starlight passing into the atmosphere is refracted toward the normal (N in the sketch), thus making the stars appear higher in the sky (closer to the zenith) than they really are.

       (b) At the zenith there is no distortion due to refraction, since the light rays are already moving along the normal and can't get bent any closer. Refraction is maximum for a star on the horizon; its position is raised nearly $\frac{1}{2}$ degree.

5-20   Any white light passing through the Earth's atmosphere will show absorption lines due to the Oxygen in the atmosphere. Thus if there are absorption lines due to Oxygen in the atmosphere of the star or planet, those lines will be masked by the Earth's Oxygen absorption lines.

5-21   The twinkling of stars is caused by the Earth's atmosphere. Since the stars are so far away, they appear as point sources of light, and in order for them to appear steady in the sky all their light should appear to come from exactly the same point. But, in fact, the photons of starlight get deflected and bumped around by the atmospheric molecules, so that the photons from a given star enter the eye from slightly different directions, and the star appears to dance around or twinkle.

5-22   Since a planet is an extended light source, the atmospheric disturbances of its light do not cause the image to appear to jump around as much as the point stellar image.

5-23   (a)   The star would twinkle. It would not be refracted. Only a little of its blue light would be scattered out.

        (b)   The star would twinkle and be dimmed. It would appear reddened because all the shorter wavelengths of starlight would have been scattered out. It would appear shifted $\frac{1}{2}$ degree toward the zenith due to refraction.

        (c)   The star would twinkle, but not be noticeably dimmed or reddened. It would be refracted 58'' toward the zenith.

5-24   The atmosphere scatters the light from the Moon, just as it scatters sunlight. The scattered moonlight reaches the observer from all parts of the sky, so that the sky takes on a milky gray color rather than appearing black. If the milky gray sky--the scattered moonlight--is brighter than the faint stars, then the faint stars become invisible to the naked eye.

5-25   City lights are allowed to shine up into the sky. The atmosphere scatters and reflects this light back down to Earth. Thus the nighttime sky in the cities appear milky gray, due to all the reflected light, and only the brightest stars are brighter than the reflected city light.

5-26   (a)   Scattered sunlight appears to reach an observer from all parts of the daytime sky, and this scattered light is brighter than the light from the stars and planets. Hence the scattered sunlight overwhelms the light from the stars and planets.

        (b)   In the mountains there is less atmosphere between the observer and the Sun, so less of the sunlight is scattered and the sky appears darker blue. Venus is the brightest planet in the sky, and it is often brighter than the dark blue mountain sky.

5-27   (1)   The yellow Sun high in the sky becomes orange and then red as it nears the horizon. This is due to the absorption of shorter wavelengths of sunlight by the atmosphere. At the zenith the sunlight passes through about 100 miles of atmosphere, and the blue part of the sunlight is absorbed and scattered. As the Sun approaches the horizon its light passes through more and more atmosphere to reach the observer, and more of the sunlight is scattered in progressively longer wavelengths. At the horizon only the very long red light gets through all the air.

5-27 (2) The other obvious change is that as the sun approaches the horizon
its observed shape changes from circular to elliptical. This is due
to differential refraction of the sunlight which pushes up the lower
parts of the Sun more than the upper parts. In the extreme, when the
Sun is really just below the horizon, refraction raises the image so
that the entire Sun appears just above the horizon.

5-28 (a) Hinders:

Scatters starlight.
Refracts starlight.
Selectively absorbs starlight.
Wind, rain and clouds hinder observations.
Twilight and refraction shorten night observing periods.

(b) Helps:

Astronomers breathe it to stay alive.
It shields astronomers and their equipment from lethal solar
   radiation and from meteorite impacts.
It retains heat at night by the greenhouse effect.

Orientation and Motions of the Earth

6-1   Cite several phenomena that prove the Earth revolves around the Sun.

6-2   Cite several pieces of evidence that indicate (but do not prove) that the Earth revolves around the Sun.

6-3   Calculate the average speed of the Earth in its orbit, in miles/second and in kilometers/second. Assume the Earth's orbit is a circle of radius one astronomical unit.

6-4   What is stellar parallax? Illustrate with a sketch.

6-5   The absence of any observations of stellar parallax was one of the arguments used by the authorities of the Inquisition against Galileo and the heliocentric theory. If the Earth revolved around the Sun, they maintained, then the nearby stars ought to shift against the background stars. Why could Galileo not observe stellar parallax?

6-6   What is stellar aberration? What two factors cause this effect?

6-7   Why do the radial velocities of some stars seem to change periodically over the course of the year?

6-8   (a)   How is the Earth's axis orientated in space? Does this orientation ever change?

      (b)   What changes are observed in the location of the north celestial pole in the sky as an observer moves south on the Earth's surface, or as he moves west on the surface?

6-9   Why does the Earth's axis precess?

6-10 (a)   What direction (relative to the horizon) do the stars appear to move in the night sky?

      (b)   Relative to the background stars, what is the apparent (observed) motion of the Sun (speed and direction)?

6-11 Suppose you wake tomorrow and find that the Earth and its rotation axis have been moved so that the equator is coincident with the ecliptic. Nothing else has changed. Which of the following would be substantially different than it is now?

6-11　(a)　The length of the mean solar day.
　　　(b)　The length of the year.
　　　(c)　The leap year system.
　　　(d)　The seasons on the Earth.
　　　(e)　The number of daylight hours in your hometown on Christmas Day.
　　　(f)　The precession of the Earth's axis.
　　　(g)　The star which serves as our north star.
　　　(h)　The heliocentric parallax of a nearby star.
　　　(i)　The behavior of a Foucault pendulum at the north pole.
　　　(j)　The oblateness of the Earth.
　　　(k)　The locations of the north, south, east and west points on your local horizon.
　　　(l)　The latitude and longitude of your home.

6-12　Suppose you wake tomorrow and find that the Earth's axis has moved so that it lies in the ecliptic plane. Nothing else has changed. Which of the above choices (a - l) would be substantially different?

6-13　Where is the center of mass of the Earth-Sun system?

6-14

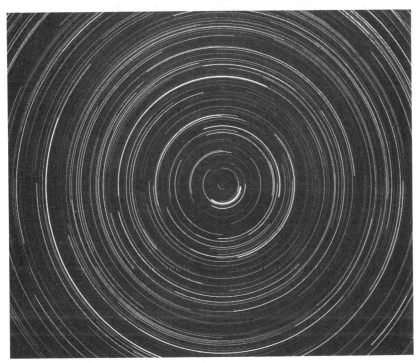

Exercise 6-14. Star trails from the Lick Observatory. (Lick Observatory photo).

The photo is a time exposure of a region of the sky.

(a)　In what direction was the telescope pointed when the photo was taken?

(b)　Approximately how many hours was the exposure time?

Exercise 6-15. Star trails near the horizon of the
Lick Observatory. (Lick Observatory photo.)

The photo is a time exposure of stars near the horizon of the Lick Observatory in California.

(a) Toward what part of the celestial sphere was the telescope pointed?

(b) Estimate the latitude of the Lick Observatory from the trails of the stars.

Seasons on the Earth

6-16 (a) What two factors contribute to causing a planet to have seasons?

(b) Which of these two factors plays the biggest role in the Earth's seasons?

6-17 How would you convince a friend that the changing Earth-Sun distance does not significantly influence our seasons?

6-18 Relatively few astronomical observations are made at observatories located in the high latitudes. Why?

6-19 Describe the seasons on the Earth if the Earth's rotation axis were

(a) Perpendicular to the ecliptic plane,

(b) Lying in the ecliptic plane.

Astronomical Coordinate Systems and the Celestial Sphere

6-20 What name describes each of the following, on the **celestial sphere**?

(a) The great circle through the north celestial pole, the zenith, and the south celestial pole.

(b) The intersection of the Earth's orbit plane with the celestial sphere.

(c) The intersection of the Earth's equatorial plane with the celestial sphere.

(d) The apparent path of the Sun on the celestial sphere.

(e) The intersection of the Earth's rotation axis with the celestial sphere.

(f) The sphere of infinite radius upon which the stars are assumed to lie.

(g) The intersections of the ecliptic with the celestial equator.

(h) The points at which the Sun is farthest from the celestial equator.

(i) The points at which the Sun is directly over the celestial equator.

6-21 (a) What are the bases of the equatorial coordinate system, the one that involves right ascension and declination?

(b) What are the bases of the horizon coordinate system, the one that involves azimuth and altitude?

6-22 How do longitude and latitude on the Earth's surface compare with right ascension and declination on the celestial sphere?

6-23 (a) What is meant by the precession of the equinoxes?

(b) What is the cause of this motion?

(c) What is the period of this motion?

6-24 (a) What is a simple method for determining the approximate latitude of a place in the northern hemisphere?

(b) Why doesn't this method work very well in the southern hemisphere?

6-25 (a) What is a circumpolar star?

(b) In general, in what part of the sky are the circumpolar stars located?

(c) Name one constellation that is circumpolar from the United States, and one from Santiago, Chile.

(d) What is the lowest declination a star can have, at your latitude, and still be circumpolar?

6-26 What is the northernmost latitude from which one can theoretically observe all the stars in the Southern Cross?

6-27 What is the noon altitude of the Sun as seen from the Aukland Public Observatory, New Zealand (latitude -37°) on

(a) 21 March,   (b) 22 June,   (c) 22 December.

6-28 From where on the Earth could each of the following be observed?

(a) The stars move parallel to the horizon.
(b) The Southern Cross is in the evening sky, and only half the stars on the celestial sphere are ever seen.
(c) The ecliptic lies on the horizon.
(d) The celestial equator passes through the zenith.
(e) The north celestial pole is at the zenith.
(f) The south celestial pole is on the horizon.
(g) The Sun rises on 23 September and doesn't set until 21 March.

6-29 At what azimuth does the Sun set after it has risen at each of the following azimuths?

(a) 80°,   (b) 90°,   (c) 115°.

6-30 (a) The right ascension of Spica is 13^h 23^m 49^s. Convert this angle into arc units.

(b) Convert the angle 187° 24' 36'' into time units.

Time and Date on the Earth

6-31 Upon what astronomical phenomenon is each of the following based?

(a) Sidereal time,
(b) Mean solar time,
(c) True solar time,
(d) Ephemeris time,
(e) Universal time.

6-32 Upon what astronomical phenomenon is each of the following time periods based?

(a) The sidereal day,
(b) The true solar day,
(c) The month,
(d) The year.

6-33 What is the approximate date (or dates) of each of the following events?

(a) The Sun rises at azimuth 90°.
(b) The vernal equinox transits the meridian at midnight.
(c) The constellation Aquarius rises at midnight.
(d) The Sun passes through the zenith of an observatory on the Tropic of Capricorn.
(e) The Sun is on the horizon for an observer at the south pole.
(f) The right ascension of the Sun is 18^h.

6-34 (a) Why do we have a leap year?

(b) If the Earth made one revolution in $365^d \ 4^h$, how often would a leap year be needed?

6-35 (a) How many sidereal days are there in one (non-leap) year?

(b) How many solar days are there in one (non-leap) year?

6-36 What is the local sidereal time when the bright star Sirius transits the meridian?

6-37 If the hour angle of the bright star Capella is $19^h \ 37^m$ at 6 PM local mean time, at what time (local mean) does Capella transit the meridian?

6-38 If Arcturus rises at 6:30 PM on 1 May, at approximately what time does it rise on 1 June?

6-39 The longitude of the observer is 90° west. The local mean time is 9:30 PM, 15 July (St. Swithin's Day). What is the Universal time and date?

*6-40 A sundial in Flagstaff, Arizona (longitude 112° west) reads 2:37 on 31 January. What is the Mountain Standard Time?

6-41 A ship is sailing from San Francisco to Tokyo. Immediately before crossing the international date line it is 9:30 AM (local standard time), 16 January. What are the time and date two minutes later, just after the date line is crossed?

6-42 Where on Earth does a given date first occur?

6-43 (a) Explain the principal of the Julian day calendar.

(b) Cite one advantage and one disadvantage of the Julian day calendar compared with the Gregorian calendar.

6-44 What is the Julian date of noon, 28 October 1978?

6-1   One proof is theoretical, involving Newton's laws. The others involve observations of the stellar reference frame.

6-2   The evidence involves observations of other bodies in the solar system, some which appear to revolve like the Earth, others whose motions suggest revolution of the Earth.

6-3   The circumference of a circle is $2\pi R$, where R is the radius. In one year the Earth travels one circumference.

       1 au = 1.496×10^8 km.   1 km = 0.6214 mi.

6-4   This is a small apparent motion of the nearby stars which is due to the Earth's revolution, shown in Exercise 11-5.

6-5   Consider the distances of the bodies involved and the sizes of the parallaxes.

6-6   Stellar aberration is due to the finite speed of both the light and the observer.

6-7   Radial velocity is the speed at which a star approaches or recedes from the Earth. This speed is observed from the Earth, which itself is moving around the Sun.

6-8   (a) Consider the effect of precession.

       (b) See Exercise 1-9.

6-9   There are three factors involved, all of which must be present in order for the body to precess. Two factors have to do with the precessing body, the other with the forces acting on it.

6-10 The directions are either east to west, or west to east.

6-11 Remember that the rotation and revolution rates of the Earth do not change,
6-12 nor does the location of anything on the surface of the Earth.

6-13 The masses of the Sun and Earth are 1.989×10^{33} gm and 5.977×10^{27} gm. The Earth-Sun distance is 1.496×10^8 km. In the sketch $m_1 d_1 = m_2 d_2$

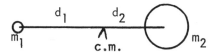

6-14 This region of the sky can be observed from anywhere in the northern hemisphere of the Earth.

6-15 (b) At the equator the stars rise perpendicular to the horizon, and at the pole they circle parallel to the horizon.

6-16   One factor has to do with the orientation of the planet's rotation axis, the other with the shape of its orbit.

6-17   Consider the dates of the seasons in both the northern and southern hemispheres. The Earth reaches perihelion in January.

6-18   Consider both the weather and the durations of sunlight and darkness in high latitude locations.

6-19   First consider what would be the seasons in part (a). Next consider how the seasons would change as the axis is moved to its present orientation, and then on to the location in part (b).

6-20   The choices are from among the vernal equinox, autumnal equinox, celestial equator, meridian, winter solstice, summer solstice, ecliptic, celestial sphere, north celestial pole, and south celestial pole.

6-21   The fundamental plane for (a) is the celestial equator, and for (b) it is the observer's horizon. Now state how the coordinates are measured relative to the fundamental planes.

6-22   They both use the same fundamental reference plane.

6-23   Consider the motion of the Earth's rotation axis, and how it affects the locations of the vernal and autumnal equinoxies.

6-24   The method involves observations of the north pole star, Polaris.

6-25   Among other characteristics, circumpolar stars appear to move in circular paths around the celestial pole.

6-26   The declinations of the four stars are

$$\alpha \; -62° \; 33' \qquad \gamma \; -55° \; 33'$$
$$\beta \; -59° \; 09' \qquad \delta \; -58° \; 12'$$

To see all four, the most southern star α Crux must be just on the southern horizon.

6-27   On the three dates the Sun is at the vernal equinox, winter solstice, and summer solstice for the New Zealand observer.

6-28   Some of them can be observed from more than one location.

6-29   The rise and set points are symmetric about the observer's meridian; e.g., an object that rises 15° east of the meridian will set 15° west of the meridian.

6-30   15° = 1 hour
        1° = 4 minutes
        15' = 1 minute
        1' = 4 seconds
        15" = 1 second

6-31   All astronomical time systems are based upon one or more motions of the Earth.

6-32   They are all based upon some motion of the Earth or Moon.

6-33   Assume the right ascension of Aquarius is 22^h. The Tropic of Capricorn is the parallel of latitude at $-23\frac{1}{2}°$.

6-34   Astronomically speaking, one year is the time required for the Earth to make exactly one revolution.

6-35   Consider the definitions of the sidereal and solar day.

6-36   The local sidereal time = hour angle of the vernal equinox. The right ascension of Sirius is $6^h 42.9^m$ (1950.0).

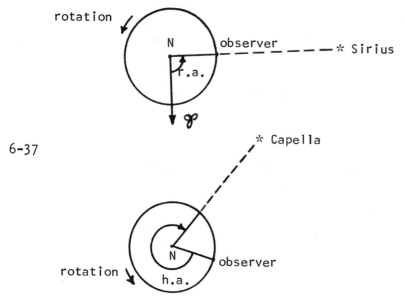

6-37

6-38   The stars rise about 4 minutes earlier each night.

6-39   The Universal time is the same as the time at the Greenwich meridian, longitude 0°.0.

6-40   You must consider both the equation of time and the longitude difference between Flagstaff and the central meridian at 105° W.

6-41   As one moves westward on the Earth, the time gets progressively earlier, so at the date line it must change to one day later.

6-42   Get a map and look around the international date line.

6-43   The Gregorian calendar is the one in use today.

6-44   The Julian date at noon 1 January 1978 is 2443510.0.

6-1   One result of Newton's law of gravity is that any two bodies revolve around their common center of mass. Since the center of mass of the Earth-Moon system is inside the Sun (see Exercise 6-13) we say that the Earth revolves around the Sun.

   The observations of stellar aberration, stellar parallax and the yearly changes in stellar radial velocities can be made only if the Earth revolves around the Sun.

6-2   (1)   Other planets are observed to revolve around the Sun.

   (2)   Satellites are observed to revolve around larger bodies.

   (3)   The seasons on the Earth.

   (4)   The change in the constellations visible in the night sky.

6-3   Circumference = 2π x 1 au = 2π x 1.496 x 10^8 km

   = 9.400 x 10^8 km.

   In one year there are $365\frac{1}{4}$ days x 24 $\frac{hr}{day}$ x 3600 $\frac{sec}{hr}$

   = 3.156 x 10^7 seconds/year.

   Speed = $\dfrac{distance}{time}$ = $\dfrac{9.400 \times 10^8 \text{ km}}{3.156 \times 10^7 \text{ sec}}$ = 29.78 km/sec.

   29.78 $\frac{km}{sec}$ x 0.6214 $\frac{mi}{km}$ = 18.51 $\frac{mi}{hr}$.

6-4   This is the apparent periodic shift in the positions of the nearby stars relative to the more distant background stars, due to the revolution of the Earth. In the figure below, the nearby star S seems to change position against the background stars, moving from S_j in January to S_m in May. The angle p is called the parallax of the star.

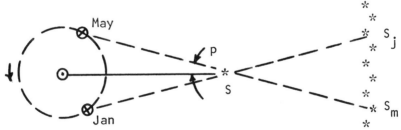

6-5   Galileo could not observe stellar parallax because the shifts are too small to be detected by the crude instruments of the 17th century. The largest stellar parallax (for the nearest known star) is only 0".76, a very small angle.

6-6   Because the Earth is moving and because light moves with a finite speed, a telescope must be moved slightly (20".5) in the direction of the Earth's motion in order for the light from a star to move directly down the axis of the telescope.

6-7   Observed stellar radial velocities change throughout the year because the observer's platform--the Earth--changes its direction of motion throughout the year. The star Ross 128 approaches the Sun at a speed of 13 km/sec, as shown below. Since the Earth's orbital speed is about 30 km/sec, Ross 128 will appear to approach the Earth at 30 + 13 = 43 km/sec December, and it will be observed to recede from Earth at 30 - 13 = 17 km/sec when observed in June.

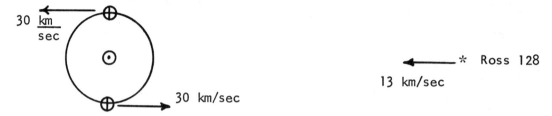

6-8   (a) The Earth's axis appears to be fixed in space, pointing to within a degree of the star Polaris and making an angle of $66\frac{1}{2}°$ with the plane of the ecliptic. The direction of the axis (but not the angle of tilt) is slowly changing due to precession.

        (b) As an observer moves south, the north celestial pole gets lower in the sky. As the observer moves west, no change is observed in the location of the pole.

6-9   The Sun and Moon exert a gravitational torque on the spinning oblate Earth, causing it to precess.

6-10   (a) The daily motion of the stars in the sky is from east to west. Like the Sun, they rise in the east and set in the west.

        (b) Relative to the stars, the Sun moves approximately one degree to the east each day.

6-11   a, b, c, i and j do not change because they depend solely on the rotation and revolution rates,

        d changes since it is dependent upon the inclination of the equator to the ecliptic.

        e changes since it is dependent upon the seasons, and there would be no seasons.

        f would change since there would be no precession.

        g would change; there would be a different pole star since the Earth's axis would have been changed in direction.

        h would not change; it depends only upon the size of the Earth's orbit.

        k and l would not change; they depend only upon surface features on the Earth.

6-12 The same choices (d, e, f, g) would be the only ones to change.

6-13

$$d_1 m_s = (1.496 \times 10^8 - d_1) m_e$$

$$d_1 = \frac{(1.496 \times 10^8) m_e}{m_s + m_e} = 1.496 \times 10^8 \times \frac{5.977 \times 10^{27}}{1.989 \times 10^{33}} = 449.6 \text{ km.}$$

6-14 (a) Toward the north celestial pole.

      (b) Approximately 8 hours.

6-15 (a) Toward the celestial equator, hence the star trails are straight lines.

      (b) The Lick Observatory is at approximate latitude 37° north, and the star trails make an angle of about 37° with the vertical in the photo.

6-16 (a) The more inclined is the equator of a planet to its orbit plane, the more extreme are its seasons.

      (b) Since the Earth-Sun distance varies by less than 3%, it is the $23\frac{1}{2}^\circ$ inclination of the equator to the ecliptic that causes the Earth's seasons.

6-17 If your friend lived in the northern hemisphere, you could simply tell him that the Earth is at perihelion in January.

Alternatively you could point out that seasonal calendar for the southern hemisphere is opposite to that for the north; whereas if the Earth-Sun distance were responsible for the seasons, then each season should occur all over the Earth at the same time.

6-18 In the winter, when the nights are very long at high latitudes, the weather is usually poor for astronomical observations. In the summer when the weather is reasonably good, the nights are very short and sometimes the sky never gets totally dark.

6-19 (a) There would be little or no seasons on Earth. Any slight changes would be due to the 3% yearly variation in the Earth-Sun distance.

      (b) The durations of the seasons would be the same as they are now, but the climate changes would be much greater. Winters would be much colder, with possibly several months of total darkness, and summers would be much warmer.

6-20 (a) meridian.
      (b) ecliptic.
      (c) celestial equator.
      (d) ecliptic.
      (e) north and south celestial poles.
      (f) celestial sphere.
      (g) vernal and autumnal equinoxes.
      (h) winter and summer solstices.
      (i) vernal and autumnal equinoxes.

6-21 (a) Right ascension is measured along the celestial equator--the fundamental plane--eastward from the vernal equinox. Declination is measured from the celestial equator, north (+) to the north celestial pole, or south (-) to the south celestial pole.

(b) Azimuth is measured along the observer's horizon--the fundamental plane--eastward from the north point. Altitude is measured from the horizon up (+) to the zenith, or down (-) to the nadir.

6-22 The two systems are very similar, except that one is attached to the rotating Earth and the other is fixed to the relatively stationary celestial sphere.

Both latitude and declination are measured north or south from the equatorial plane, and both longitude and right ascension are measured along the equatorial plane.

An important difference is that r.a. is measured eastward from the vernal equinox; whereas longitude is measured east or west from the Greenwich meridian.

6-23 (a) The locations of the equinoxes on the celestial sphere are slowly changing. They are moving westward along the ecliptic at the rate of about 52" per year.

(b) The cause of this motion is the precession of the Earth's axis. Because the axis is slowly moving, the equator is also moving. And because the equator is moving, so is the line of intersection of the equator and ecliptic. The equinoxes are the points where this moving line intersects the celestial sphere.

(c) The period is about 26,000 years.

6-24 (a) The altitude of the bright north pole star, Polaris, is equal to the latitude of the observer, to within one degree.

(b) The method doesn't work so well in the southern hemisphere because there is no bright star in the region of the south celestial pole.

6-25 (a) A star that never goes below the horizon.

(b) They are found in the polar regions of the sky, near the celestial poles.

(c) Ursa Minor, the Little Dipper or the Little Bear, is the constellation closest to the north celestial pole, so it is circumpolar for most northern hemisphere observers. Cepheus, Cassiopeia, and Camelopardalis might be circumpolar, depending upon the observer's latitude.

The constellations closest to the south celestial pole are Octans, Chamaeleon, Apus and Mensa.

(d) The distance from the celestial pole to the horizon is equal to the observer's latitude ϕ. Thus any star within ϕ degrees of the pole is circumpolar, or any star with declination greater than $90° - \phi$ is circumpolar.

6-26 $\phi = 90° - 62° 33'$

$\qquad = +27° 27'.$

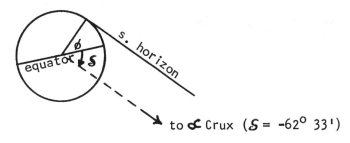

to α Crux ($\delta = -62° 33'$)

6-27 (a) $90° - 37° = 53°$ (above the northern horizon).

(b) $90° - 37° - 23° = 30°$ "   "

(c) $90° - 37° + 23° = 76°$ "   "

6-28 (a) At either the north or south pole.
(b) At the south pole.
(c) On the arctic or antarctic circles.
(d) On the equator.
(e) At the north pole.
(f) On the equator.
(g) At the south pole.

6-29 (a) $360° - 80° = 280°.$
(b) $360° - 90° = 270°.$
(c) $360° - 115° = 245°.$

6-30 (a) $13^h = 52°$

$23^m = 5° \ 45'$

$49^s = \underline{\quad 12' \ 15''}$
$\qquad \quad 57° \ 57' \ 15''$

(b) $187° = 12^h \ 28^m \ 00^s$

$24' = \qquad \quad 1^m \ 36^s$

$36'' = \underline{\qquad \qquad 2\overset{s}{.}4}$
$\qquad \quad 12^h \ 29^m \ 38\overset{s}{.}4$

6-31 (a) The rotation of the Earth with respect to the stars.

(b) The rotation of the Earth with respect to the fictitious mean sun.

(c) The rotation of the Earth with respect to the true Sun.

(d) The revolution of the Earth with respect to the stars.

(e) This is based upon the same motion as is mean solar time, except that noon occurs when the mean sun crosses the Greenwich meridian.

6-32 (a) One rotation of the Earth with respect to the vernal equinox.

(b) One rotation of the Earth with respect to the true Sun.

(c) One revolution of the Moon around the Earth (approximately).

(d) One revolution of the Earth.

6-33 (a) When the Sun is at an equinox: 21 March or 23 September.

(b) When the Sun is at the autumnal equinox: 23 September.

6-33 (c) When Aquarius is in opposition, or when the right ascension of the Earth is 22^h, which is one month before the Sun is at the autumnal equinox, which is 23 August.

      (d) When the Sun is at its farthest point south, at the winter solstice (for northern observers), 22 December.

      (e) When the Sun is at either equinox: 21 March or 23 September.

      (f) Since the Sun is at 0^h when at the vernal equinox, then it is at 18^h three quarters of a year later, 22 December.

6-34 (a) Leap year is needed because the Earth does not make exactly one revolution in an integral number of days. Rather, it requires $365\frac{1}{4}$ days to make one revolution. So that the year by which we live will always begin at midnight and have an integral number of days, we have three 365-day years and then one 366-day year every four years. (Alternatively, my friend says the reason is so girls can chase boys.)

      (b) Every 6 years.

6-35 (a) 365 days.

      (b) 366 days.

6-36 Local sidereal time = hour angle of the vernal equinox

                           = right ascension of an object on the meridian.

                           = $6^h 42.9^m$

6-37 Time until transit = $24^h - 19^h 37^m = 4^h 23^m$.

      Transit time = 6 PM + $4^h 23^m$ = 10:23 PM.

6-38 Since there are 31 days between 1 May and 1 June, the star will rise $4 \times 31 = 124$ minutes = 2 hours 4 minutes earlier = 4:26 PM.

6-39 Every 15° west of Greenwich meridian is one hour earlier than Greenwich mean time (which is Universal time). So the time zone centered on the 15° meridian is one hour earlier, the one centered on the 30° meridian is two hours earlier, and so on. The 90° meridian is six hours earlier. So if it is 9:30 PM at the 90° meridian, it is six hours _later_ at the Greenwich meridian or 3:30 the next day, 16 July.

6-40 From the equation of time of 31 January.
      Apparent time - mean time = -13 minutes.
      Thus the local mean solar time at Flagstaff = 2:37 - 0:13 = 2:24 PM.
      (PM because the sundial doesn't work at 2:37 AM.)

      Mountain Standard Time (MST) is the mean solar time at the central meridian (longitude 105° W), where it is later than at Flagstaff. Since each degree of longitude = four minutes of time, $112^o - 105^o = 7^o$ = 28 minutes.

      Thus MST = 2:24 + 0:28 = 2:52 PM.

6-41 9:32 AM, 17 January.

6-42 At those locations immediately west of the international date line, such as Tonga Island in the South Pacific.

6-43 (a) Each day is identified by one number. The days are numbered consecutively, starting at noon 1 January 4713 BC, which is JD 0.0.

(b) The advantage of the system is that each day is identified by only one number; whereas in the Gregorian system each day requires three quantities: year, month, and day.

The disadvantage is that is is very difficult to relate events in the two systems, Julian and Gregorian.

6-44 JD 2443810.0.

7 THE MOON

Appearance in the Sky: Phases and Eclipses

7-1 What is the phase of the Moon during each of the following?

(a) The Moon is in conjunction with the Sun.
(b) The Moon is in opposition to the Sun.
(c) The Moon rises at sunset.
(d) The Moon sets at sunset.
(e) The Moon rises at noon.
(f) The Moon sets at noon.
(g) The Moon is observed above the western horizon just after sunset.
(h) The Moon sets just before sunrise.
(i) The Moon is on the meridian at sunset.
(j) A total lunar eclipse.
(k) A total solar eclipse.
(l) A partial lunar eclipse.
(m) An annular solar eclipse.

7-2 (a) During which lunar phase does one observe the most earthshine on the Moon?

(b) Which would appear brighter: the full Moon as seen from Earth, or the full Earth as seen from the Moon?

7-3 What is the relationship between the lunar phases as seen from Earth and the Earth's phases as seen from the Moon?

7-4 Fill in the table to indicate which of the seven phenomena can be seen by the naked eye from the five locations:

	Earth at noon	Earth at midnight	Moon near-side noon	Moon near-side midnite	Moon far-side noon
Stars & planets					
Dark sky					
Crescent Earth					
The Moon					
Total solar eclipse					
Jupiter on the midnight meridian					
Annular solar eclipse					

Exercise 7-5. Venus and waxing crescent Moon. (Photo by Ray Martin and Larry Barstow).

The photo is Venus and the waxing crescent Moon taken in the early evening of 16 April 72. On the sketch below (**not** to scale) show the locations of

(a) The observer's horizon,
(b) The Moon,
(c) Venus.

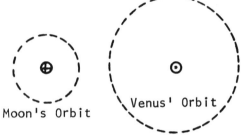

Moon's Orbit   Venus' Orbit

7-6 (a) What is the angular size of the Moon as observed from Earth?

(b) What is the angular size of the Earth as observed from the Moon?

(c) What is the angular size of the Sun as observed from either?

7-7 Why does the Moon appear so much larger near the horizon than it does when it's overhead?

7-8 How far away would you have to hold a half-dollar in order for it to have the same angular size as the Moon--to just cover it.

7-9 What ground-based observations indicate that

(a) there is no atmosphere on the Moon?

(b) the lunar maria are not water oceans?

7-10 (a) Describe the differences between total, annular and partial eclipses of the Sun.

(b) Why is it not totally dark during a total solar eclipse?

7-11 Total lunar eclipse or total eclipse:
 (a) Which lasts longer? Why?
 (b) Which occurs more frequently?
 (c) Which is one more likely to observe? Why?

7-12 (a) Under what conditions does a lunar or a solar eclipse occur?
 (b) Why doesn't an eclipse occur every two weeks?

Lunar Surface Features and Environment

7-13 Which of the following could be observed by a daytime observer on the near-side of the Moon?

 (a) Noise pollution.
 (b) Crescent Earth.
 (c) Total solar eclipse.
 (d) Annular solar eclipse.
 (e) Water pollution.
 (f) Air pollution (on the Moon).
 (g) Retrograde planetary motion.
 (h) Clouds on Earth.
 (i) Visible light.
 (j) Stars in the sky.
 (k) Radio waves from Earth.

7-14

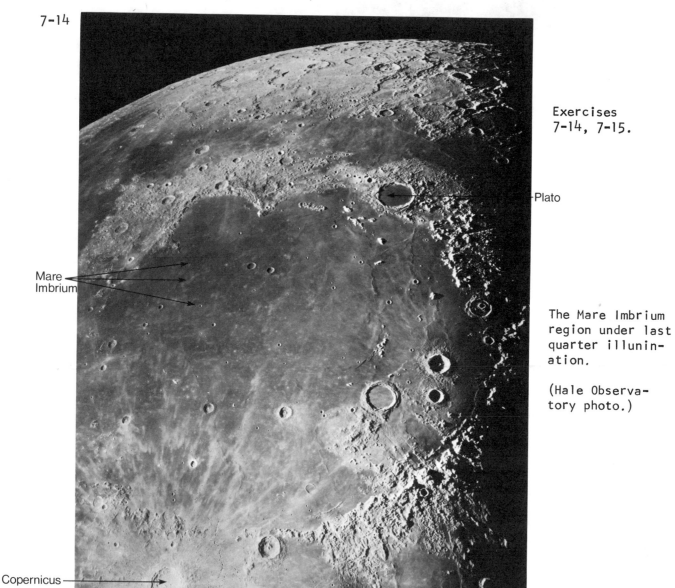

Exercises
7-14, 7-15.

Plato

Mare Imbrium

The Mare Imbrium region under last quarter illunination.

(Hale Observatory photo.)

Copernicus

7-14 Explain why the lunar features (craters, mountains) seem more prominent along the terminator in the photo than they do away from the terminator.

7-15 List the following features on the photo in order of their probable occurrence, oldest first.

    (a) Crater Copernicus,
    (b) Rays from Copernicus,
    (c) Mare Imbrium,
    (d) Crater Plato.

7-16 (a) What proportion of the near side of the Moon is covered with maria?

    (b) What proportion of the far side of the Moon is covered with maria?

7-17 How are the ages of lunar rock samples determined?

7-18 What evidence of water has been found on the Moon?

7-19 Why does the Moon have very little atmosphere?

7-20 You are camped in Sinus Medii at the center of the Moon's near side. The Moon is at first quarter as seen from Earth.

    (a) Where is the terminator, relative to your camp?

    (b) Where do you see the Sun and the Earth?

    (c) In how many days will you see the Sun set, the Earth set?

7-21 It is believed that some of the lunar maria may have been caused by the impact of small asteroids, and that these events occurred after the Moon was covered with craters. Cite two pieces of evidence--two known or observed phenomena about the Moon--that support this idea.

7-22 (a) Cite two arguments that support the impact origin of the lunar craters.

    (b) Cite two arguments that support the volcanic origin of the lunar craters.

7-23 (a) What are lunar mascons?

    (b) Where on the Moon are the mascons located?

    (c) How were the mascons discovered?

7-24 What are some basic differences in the surface environments of the Earth and Moon that are due to the absence of atmosphere on the Moon?

7-25 Compared with an Earth-based telescope, what is one advantage and one disadvantage of having an optical telescope on the Moon?

Lunar Motions, Gravity and the Tides

7-26 What three bodies are primarily responsible for ocean tides?

7-27 (a) In which direction does the Moon rotate, clockwise or counter-clockwise?

7-27 (b) Why does the Moon keep the same face toward the Earth?

7-28 What is the primary influence of the Earth on the Moon?

7-29 (a) Define the synodic period of the Moon and give its duration.

(b) Define the sidereal period of the Moon and give its duration.

7-30 If the Moon's orbit were retrograde (east to west) rather than direct (west to east), would the synodic month be longer or shorter than the sidereal month?

7-31 (a) During which lunar phases do we experience the highest high tides and the lowest low tides?

(b) If high tide at San Diego occurs at 9 AM on Saturday, when does the next low tide occur and when does the next high tide occur?

7-32 If an astronaut weighs 180 lbs on Earth, what would be his weight on the lunar surface?

7-33 (a) The position of the full moon in the sky is noted. Then how far away (how many degrees) will be the position of the next full moon?

(b) Through how many zodiacal constellations is this apparent motion?

*7-34 Approximately how long do radar waves take to reach the Moon and return to Earth?

*7-35 Where is the barycenter of the Earth-Moon system located?

*7-36 Where is the gravitational balance point between the Earth and the Moon-- the point where a spacecraft would feel equal gravitational attractions from each body?

HINTS TO EXERCISES ON THE MOON

7-1   Review the meanings of conjunction and opposition, and how to tell time from the position of the Sun relative to the observer.

7-2   Earthshine is sunlight that is reflected first off the Earth and then off the Moon.

7-3   The Earth goes through the same cycle of phases as does the Moon, but the Earth and Moon are never simultaneously in the same phase.

7-4   Remember that the Moon has no atmosphere. The Earth's diameter is about 4 times larger than the Moon's, and the angular diameters of the Moon and the Sun (seen from Earth) are both about $\frac{1}{2}^{\circ}$.

7-5   In the evening the Sun is just below the horizon while the Moon and Venus are just above it.

7-6   Same hint as 7-4.

7-7   In reality the Moon's observed size is the same in both locations.

7-8   This can be done by experiment or calculation.

7-9   Some clues can be obtained from a picture of the Earth from space.

7-10   In the three cases: the central part of the Sun is covered, all of the Sun is covered, and part of the Sun is covered by the Moon.

7-11   (a) Consider the direction of motion of the Moon and the shadows of the Earth and Moon.

       (c) Consider from where each can be observed.

7-12   Consider the orientation of the Moon's orbit plane relative to the ecliptic plane.

7-13   The Moon has no atmosphere and no water.

7-14   Consider how the features are illuminated and shaded.

7-15   Older features are sometimes destroyed by newer features.

7-16   Look on a map of the Moon. There is much less maria on the far side of the Moon than on the near side.

7-17   Over long periods of time certain elements spontaneously change into other elements.

7-18   Any evidence would have been in the lunar rock samples.

7-19   Consider what property of a moon or planet enables it to retain an atmosphere. See Exercises 5-14 and 8-27.

7-20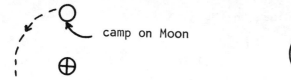

camp on Moon

Sun

7-21 Consider in what ways the circular maria resemble lunar features that are known to be of impact origin.

7-22 For these kind of arguments the Earth is often used as a well-understood analogy.

7-23 The word "mascon" stands for mass concentration.

7-24 Consider the ways in which the atmosphere affects the Earth's surface.

7-25 Refer to Exercises 4-40 and 4-41.

7-26 The tides are caused primarily by the gravitational attractions of two other members of the solar system.

7-27 Both the Earth's rotation and the Moon's revolution are counter-clockwise as seen from above the north pole.

7-28 It is gravitational.

7-29 One period uses the Sun as a reference; the other uses the stars.

7-30 See Exercise 7-29.

7-31 The Earth makes one rotation relative to the moving Moon in $24^h\ 50^m$.

7-32 The Earth's surface gravity is 6 times greater than the Moon's.

7-33 The Moon's synodic period is given in Exercise 7-29. Each of the zodiacal constellations occupy about 30 degrees of the ecliptic.

7-34 Radar waves travel at the speed of light (300,000 km/sec), and the Moon is about 384,000 km from the Earth.

7-35 Assume the Earth-Moon distance is 384,000 km., and the mass of the Earth
7-36 is 81.4 times larger than the Moon's mass.

7-1 (a) New
     (b) Full
     (c) Full
     (d) New
     (e) First quarter

     (f) Last quarter
     (g) Waxing crescent
     (h) Waxing gibbous
     (i) First quarter

     (j) Full
     (k) New
     (l) Full
     (m) New

7-2 (a) Crescent.
     (b) The full Earth is brighter because (1) it is larger, and (2) it has
         a greater albedo than the Moon.

7-3
Earth	Moon
New	Full
First quarter	Last quarter
Full	New
Last quarter	First quarter
New	Full

7-4
Stars & planets	X	X	X	X
Dark sky	X	X	X	X
Crescent Earth				
Moon	X X	X	X	X X
Total solar ecl.	X	X		
Jupiter midn. mer.	X		X	
Annular solar ecl.	X			

7-5

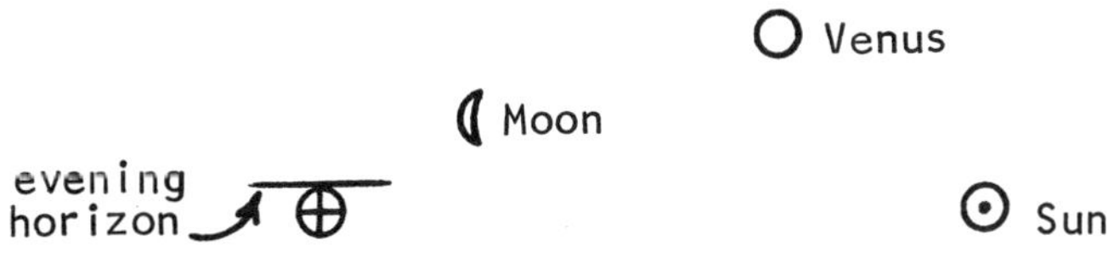

O Venus

☾ Moon

evening
horizon ↗ ⊕   ☉ Sun

7-6 (a) $\frac{1}{2}°$ (b) 2°   (c) $\frac{1}{2}°$

7-7 The reason is not known for certain, but it seems to be psychological. The
     Moon seems closer (and thus larger) when it is seen next to relatively
     nearby objects (trees, buildings, etc.) on the horizon, and farther when
     seen alone high in the sky.

7-8 The half dollar has to be held at 3.4 meters (a dime at 2 meters).

7-9 (a) Neither clouds nor twilight zones are observed on the Moon. When the
         Moon occults a star, the star's light goes out sharply. The spectrum
         of moonlight shows no evidence of absorption due to a lunar atmosphere.

7-9   (b)   No clouds are observed over or near the maria. The maria are poor
              reflectors of sunlight; whereas water is a good reflector of light.
              It is known that in the absence of atmosphere (Exercise 7-9 (a))
              liquid water evaporates.

7-10   (a)   During a total eclipse the Moon is close enough to the Earth that its
              shadow reaches the Earth's surface. Anyone inside the shadow sees the
              Sun totally covered by the Moon.

              In an annular eclipse the Moon is so far from the Earth that its shadow
              does not reach the Earth's surface. An observer near the Sun-Moon line
              sees a ring of sun around the smaller Moon.

              In a partial eclipse the observer is offset from the Sun-Moon line and
              sees only part of the Sun obscurred by the Moon.

       (b)   The sunlight falling on the atmosphere outside the eclipse shadow is
              scattered into the shadow region.

7-11   (a)   Lunar eclipse, because the Moon is moving in the same direction in
              space as is the Earth's eclipsing shadow.

       (b)   Solar eclipses are more frequent than lunar eclipses.

       (c)   The total lunar eclipse is more likely to be observed because it can
              be seen from nearly half the Earth (the half facing the Moon); whereas
              the total solar eclipse is observable only from within the narrow
              eclipse shadow band.

7-12   (a)   When the Moon and Sun are simultaneously near the nodes of the Moon's
              orbit.

       (b)   Since the Moon's orbit is inclined to the ecliptic, and Moon is
              usually above or below the Earth-Sun line when in conjunction or
              opposition.

7-13   b, c, g, h, i, j, k

7-14   Features near the terminator stand out because there is contrast between
       their illuminated and shaded parts.

7-15   d (oldest), c, a & b.

7-16   (a)   About 50%.

       (b)   Less than 10%.

7-17   By the method of radioactive dating.

7-18   None. The lunar rock samples strongly indicate that there has never been
       water on the Moon.

7-19   The primary factor is that the Moon has too little surface gravity to retain
       light atmospheric gas molecules. Also, at the Moon's relatively close
       proximity to the Sun, any atmospheric molecules receive solar energy that
       helps them escape the lunar gravity.

7-20 (a) The terminator runs through the camp.

  (b) The Sun is just rising; the Earth is at the zenith.

  (c) 14 days, never.

7-21 Some of the maria are round and surrounded by mountains (Imbrium) like the round craters are surrounded by their walls. The mountains are steepest on the side next to the maria, like the crater walls. Partially destroyed craters near the edges of the maria appear to be older than the maria which destroyed them.

7-22 (a) Since meteorites are known to fall on Earth and cause impact craters, it follows that they could do the same on the Moon. Impact craters have been found in lunar rock samples. Lunar rays look like they were caused by material ejected by an impact.

  (b) Small craters on the walls of large craters are common features on the Moon and they are common volcanic features on Earth. So many of them on the Moon is very unlikely from random impacts. Similarly, linear chains of craters are unlikely to result from random impacts, but they are common volcanic features. Craters on tops of peaks are common volcanic features, and a few are found on the Moon. They are unlikely random impact features.

7-23 On the Moon there is no air or noise pollution, no wind or water erosion, and no signs of life. The Earth abounds with all of these. On the Moon are many small craters that were caused by impacting meteorites which would have been burned up by the atmosphere had they struck the Earth.

7-24 (a) Large regions on the Moon where the material seems unusually dense.

  (b) In or beneath the lunar maria.

  (c) From studies of the gravitational attractions acting on the Lunar Orbiter spacecraft.

7-25 See the answer to Exercise 4-40.

7-26 The Sun, Moon and Earth.

7-27 (a) Counterclockwise, like the Earth.
  (b) Because its rotation period equals its revolution (around the Earth) period.

7-28 The Earth's gravity keeps the Moon in orbit around the Earth.

7-29 (a) One revolution of the Moon with respect to the Sun, or the period of the lunar phase cycle: $29\frac{1}{2}$ days.

  (b) One revolution with respect to the stars: 27 1/3 days.

7-30 Shorter.

7-31 (a) Full and new.

  (b) 9 AM + 6^h $12\frac{1}{2}^m$ = 3:12 PM.
      9 AM + 12^h 25^m = 9:25 PM.

7-32   30 lbs.

7-33   (a)   Since the Earth moves 360° in $365\frac{1}{4}$ days, in $29\frac{1}{2}$ days (one lunar phase cycle) the Earth carries the Moon through 29°.

        (b)   Approximately one zodiacal constellation during a phase cycle.

7-34   Time = $\dfrac{\text{distance}}{\text{speed}}$ = $\dfrac{2 \times 384,000 \text{ km}}{300,000 \text{ km/sec}}$ = 2.56 sec.

7-35

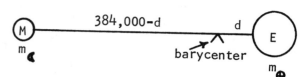

$m_{\oplus} d = m_{\mathbb{C}} (384,000-d)$

$d = 384,000 \dfrac{m_{\mathbb{C}}}{m_{\oplus}+m_{\mathbb{C}}} = 384,000 \dfrac{1}{\dfrac{m_{\oplus}}{m_{\mathbb{C}}}+1} = 384,000 \dfrac{1}{82.4} = 4660$ km

7-36

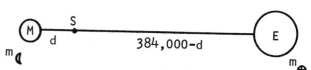

At point S (at a distance d from the Moon) the gravitational attractions are:

Due to $\oplus$: $F_{\oplus} = G \dfrac{mm_{\otimes}}{(384,000-d)^2}$

$\begin{bmatrix} m \text{ is the mass of the} \\ \text{spacecraft at S.} \end{bmatrix}$

Due to $\mathbb{C}$: $F_{\mathbb{C}} = G \dfrac{mm_{\mathbb{C}}}{d_2}$

If the forces are equal, then

$F_{\oplus} = F_{\mathbb{C}}$: $\dfrac{m_{\oplus}}{(384,000-d)^2} = \dfrac{m_{\mathbb{C}}}{d^2}$,   G and m cancel out

$\left(\dfrac{m_{\oplus}}{m_{\mathbb{C}}} -1\right) d^2 + 768,000 \; d - 384,000^2 = 0$

$d = 39,000$ km

Planetary Observations and Appearances in the Sky

8-1   (a)   Which of the nine planets are visible to the naked eye?

(b)   Which appears brightest in the sky?

(c)   Which planets are easily visible now in the evening sky, in the morning sky?

8-2   (1)   (2)   (3)

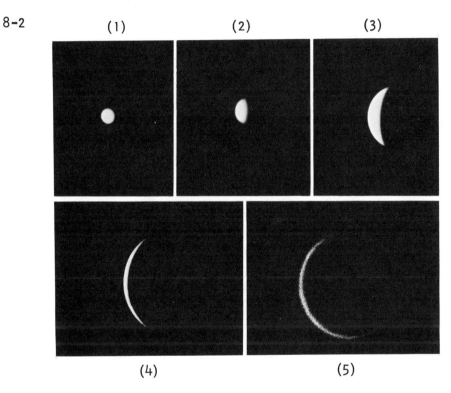

(4)   (5)

Exercise 8-2.
Five photos of
Venus all in the
same scale.
(Lowell Observatory photograph.)

The five photos of Venus are all at the same scale.

(a)   In which photo is Venus closest to Earth, and why?

(b)   In which photo is Venus at elongation?

(c)   What in the photos indicates that Venus has a dense atmosphere?

8-3   List several different ways to distinguish planets from stars in the night sky.   (There are at least six.)

8-4   Why are the planets always observed in the sky near the ecliptic?

8-5   If a bright light appeared in the sky in the region of the Big Dipper, what would you conclude about the possibility that it was one of the known planets? Give your reasoning.

8-6   (a)   Which planets are observed to go through all phases (like the Moon)?

       (b)   What is the phase of Mercury when it is at greatest elongation?

       (c)   What is the phase of Jupiter when it is at opposition?

8-7   (a)   What factors determine how bright a planet appears in the sky?

       (b)   For each of the factors in (a), which planet is the brightest considering that factor alone?

8-8   (a)   Why do the apparent diameters (as observed with a telescope) and brightnesses of the planets change?

       (b)   In which planet would you observe the greatest percentage change in its apparent diameter and brightness? Explain.

8-9   (a)   What are the best times and locations in the sky for observations of the inferior planets?

       (b)   What are the best times and locations in the sky for observations of the superior planets? Explain your answer.

8-10   Which of the planets can never be observed at opposition?

8-11   When Saturn is at opposition, at approximately what time does it

       (a)   Rise,   (b) Cross the meridian,   (c) Set?

8-12   Why can planets sometimes be observed when they are at superior conjunction, and why (same reason) do the inferior planets not always transit the Sun when they are at inferior conjunction?

8-13   For most astronomical observations, planets are observed when they are farthest from the Sun as seen in the sky.

       (a)   Name the two configurations (one for each type of planet) at which planets are at the greatest angular distance from the Sun.

       (b)   For each of the configurations named in (a), state which of the nine planets can be observed at that configuration.

8-14   Approximately how far (how many degrees) does Jupiter move on the celestial sphere in one year?

8-15   (a)   Describe retrograde planetary motion.

       (b)   Near what configurations do the inferior and superior planets retrograde?

8-16 When Venus and Mercury are "evening stars"--seen just after sunset--are they east or west of the Sun? Use a diagram to explain.

*8-17 On 19 January 1979 the planet Venus will be at greatest western elongation. If the Sun rises at 7:06 on that day, at approximately what time will Venus rise?

*8-18 (a) What are the two conditions that determine when during the year (which weeks or month) Mercury is best situated for telescopic observation? Assume Mercury's orbit is circular in the ecliptic.

Planetary Surface Environments and Features

8-19 List one unique feature about each planet.

8-20 Why would the planet Saturn be an unpleasant place to live? List several reasons.

8-21 Why would the planet Mercury be an unpleasant place to live? List several reasons.

8-22 Why was Mars selected as the target planet for the Viking spacecraft?

8-23 (a) Cite two pieces of evidence (or two kinds of observations) that show that the rings of Saturn are not solid objects.

(b) What are two hypotheses for the origin of Saturn's rings?

8-24

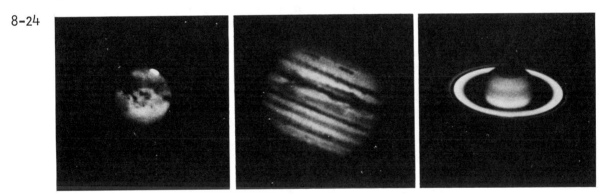

Exercise 8-24. Three planets. (Lick Observatory photo.)

(a) Identify the three planets in the photos.

(b) What two observable features in the photo of Jupiter are a result of the planet's relatively rapid rotation?

8-25 Describe the seasonal changes on Mars that can be observed telescopically from Earth?

8-26

Exercise 8-26.
A cratered region
of Mars from
Mariner 6.
(NASA-Jet Pro-
pulsion Lab photo.)

The photo is of the Hellespontus region of Mars, taken by Mariner 6.

(a) In what way do the Martian craters appear different from the craters on the Moon?

(b) Explain how the Martian craters got their different appearance, assuming they were formed by the same mechanism as the lunar craters.

(c) What does the appearance of the Martian craters indicate about the existence of a Martian atmosphere?

8-27 List the factors that determine whether a planet is able to retain an atmosphere. (There are at least three.)

8-28 List the characteristics of a planet which determine the maximum and minimum temperatures that occur on its surface. (There are at least four.)

8-29 Mariner 10 has discovered many craters on the surface of Mercury. Do you expect that many craters will also be found on the surface of Venus? Explain your answer.

*8-30 Calculate your weight if you stood on the surface of Saturn (assuming that is possible, which it isn't).

8-31 (a) Cite two pieces of evidence which indicate that there has been sub-stantial amounts of liquid on the Martian surface in the not-to-distant past.

      (b) Why is it believed that Mars has no liquid water on its surface now?

General Characteristics of the Planets

8-32   There are two ways that planets are often categorized.

   (a)   Categorize the planets according to location in the solar system, stating the characteristics and members of each category.

   (b)   Categorize the planets according to physical characteristics, stating the characteristics and members of each category.

8-33   What is the fundamental difference between stars and planets?

8-34   In order for a theory of the evolution of the planetary system to be acceptable to astronomers, the theory must be consistent with the observed solar system. In other words, the theory must be able to explain the general features of the planetary system as we observe it today. What are the important characteristics--both physical and orbital--shared by all the planets?

*8-35   At the November 1976 opposition, Jupiter's angular polar diameter was $45\overset{''}{.}69$ and its equatorial diameter was $48\overset{''}{.}95$.

   (a)   Compute the oblateness of Jupiter.

   (b)   In November 1976 Jupiter's opposition distance was 4.023 au. Compute Jupiter's equatorial diameter in kilometers.

*8-36   Calculate the mass of Jupiter using the motion of its satellite I, named Io. The sidereal period of Io is 1.769 days, and it is 421,600 km from the center of Jupiter.

Planetary Formation and Discovery

8-37   How was the planet Uranus discovered?

8-38   How was the planet Neptune discovered?

8-39   How was the planet Pluto discovered?

8-40   (a)   How was "slow moving object 1977 Kowal" (Chiron) discovered?

   (b)   Why was it difficult to decide how to classify this object?

8-41

Exercise 8-41. Two photos of the planet Pluto. (Lick Observatory photo.)

These are two photos of the same part of the sky taken 24 hours apart. Locate the planet Pluto.

8-42   Describe any attempts at the prediction and/or discovery of planets since the discovery of Pluto.

8-43   (a)   List several possible ways that the planetary system might have been formed.

(b)   For each of your ideas in (a), list one idea or piece of evidence that supports the idea.

(c)   For each of your ideas in (a), list one idea or piece of evidence that contradicts the idea.

Planetary Orbits and Motions

8-44   Although Venus' rotation rate was not determined until the 1960's, it was suspected that the planet rotated very slowly. Cite two things about Venus' appearance (telescopic, optical) that indicate it is a slow rotator.

8-45   List two ways that have been used to determine the rotation speeds of planets. For each method, list one planet upon which the method has been used.

*8-46 (a) Define the synodic and sidereal periods of a planet.

(b) Given that the sidereal period of Jupiter is 12 years, calculate Jupiter's synodic period in years.

(c) Given that the synodic period of Venus is 1.6 years, calculate Venus' sidereal period in years and in days.

*8-47 Determine which planet has the largest perihelion distance, and calculate that distance in au's and in kilometers.

*8-48 Between what limits (degrees) does the maximum elongation of Mercury vary? Assume the Earth is in a circular orbit of radius 1 au.

*8-49 (a) Which planet comes closest to the Earth in opposition?

(b) Assuming the planetary orbits are circles, what is the opposition distance (in au's) of the planet in (a).

(c) Using elliptic planetary orbits, calculate the closest approach distance (in au's) of the planet in (a).

*8-50 Show that the planet Uranus obeys Kepler's third law.

8-1   Those planets that are not visible to the naked eye were discovered only in the last 200 years. Remember the Earth is a planet.

8-2   Venus is farthest from Earth in photo number one.

8-3   The planets appear different from stars in brightness, in location in the sky, and in relative motion. See Exercise 1-15.

8-4   The ecliptic is the plane of the Earth's orbit, and it is also the apparent path of the Sun on the celestial sphere.

8-5   Consider the locations in the sky of the Big Dipper and the regions where the planets move.

8-6   Mercury at max. elongation

8-7   The factors fall into two classes: those dealing with the distances of the bodies involved, and those that determine how much light the planet reflects. There are five factors.

8-8   Both factors depend (among other things) upon the distance from the observer to the planet.

8-9   Planets are best observed against a dark background, and when their light passes through a minimum amount of the Earth's atmosphere.

8-10   ⊙   ⊕ ——————→   opposition

8-11   ⊙   ⊗   ——O— Saturn at opposition

8-12   Consider the inclinations of the planetary orbits.

8-13   One configuration is for the inferior planets and one is for the superior planets.

8-14   Jupiter's sidereal period is 11.87 years, during which time it moves through 360 degrees on the celestial sphere.

8-15   As you overtake a car moving in the same direction, the other car appears to move backwards (relative to distant background objects) as you get close to it and pass it.

8-16 Figure out from the diagram which side of the Earth is the sunset side. Then figure out where an inferior planet would have to be in its orbit to be observed at sunset.

west of Sun

☉

east of Sun

⊗↗   direction of Earth's rotation

8-17 Venus will be 46° west of the Sun, and the Earth rotates 360° every 24 hours, or one degree every four minutes.

Venus
○
＼
  ＼
   ＼
46° ＼
      ⊕↗   direction of Earth's rotation
☉＿＿＿＿＿

8-18 One condition is related to Mercury's position in its orbit relative to the Earth, its configuration.

The other condition deals with the relationship between the ecliptic and the observer's horizon.

Remember too, Mercury is best observed when it is high above the horizon and against a dark sky.

8-19 The unique features may deal with physical characteristics, surface environment, characteristics of the orbits, or history.

8-20 These two questions deal entirely with the hostile (to man) environments
8-21 that would be found on the planetary surfaces.

8-22 A primary goal of Viking is the search for extraterrestrial life forms.

8-23 We cannot see through solid natural objects.

8-24 All three are easily visible to the naked eye.

8-25 There are changes in both the Martian polar caps and the coloration.

8-26 Mars has an atmosphere and high winds.

8-27 A planet loses its atmosphere when all the atmospheric molecules are able to escape the planet's gravitational attraction. See Exercise 7-19.

8-28 Three of the factors involve the environment and physical characteristics of the planet. The fourth factor involves the planet's location in the solar system.

8-29 Mercury has very little atmosphere, whereas Venus has a very dense atmosphere. Also, many of the craters on Mercury are probably impact craters which have changed very little since they were formed.

8-30   You must use Newton's law of gravity, on the Earth and on Saturn. Divide the two equations to eliminate the constants.

$$F = G \frac{m_1 m_2}{r^2}$$

8-31   (a)   The evidence was found by Mariner 9.

8-32   (a)   The categories are called inferior and superior and are relative to the Earth.

       (b)   The categories are called terrestrial and jovian, and are named for typical members.

8-33   The answer deals with their energy sources.

8-34   In what ways are the planets the same physically, and in what ways are their orbits the same or similar?

8-35   Oblateness is defined as $\dfrac{\text{equatorial diameter} - \text{polar diameter}}{\text{equatorial diameter}}$.

       If the linear diameter and the distance are in the same units, then

       angular diameter (radians) = $\dfrac{\text{linear diameter}}{\text{distance}}$.

8-36   Use Newton's modification of Kepler's third law

       $P^2 (m_1 + m_2) = a^3$, where

       P is the period of the satellite in years,

       a is the semi-major axis of the satellite orbit in au's,

       m_1 is the mass of Jupiter in solar masses,

       m_2 is the mass of Io in solar masses, which is so small that it is ignored in this problem.

8-37   Uranus was the first planet to be discovered, the closest planet that was not known to the ancients.

8-38   Both mathematics and observations of other planets were required.

8-39   A very long search was required.

8-40   The object is a few hundred kilometers in diameter, and its orbit extends from about 1 au inside Saturn's orbit, nearly out to Uranus' orbit.

8-41   It is the object which has changed position.

8-42   See Tombaugh's search after he found Pluto.

       See the prediction of Brady and the subsequent search and mathematical analyses in the early 1970's.

8-43 A successful hypothesis must be consistent with the major known features of the solar system. See Exercise 8-34.

8-44 Compare the appearances of the fast rotators, Jupiter and Saturn, with the appearance of Venus. Some of the differences are due to the great differences in their rotation rates.

8-45 One method has been applied to planets with visible surface features.

One method involves radio signals sent from Earth.

Two other methods involve the analysis of the light received from the planets.

8-46 The relationships between sidereal and synodic periods are,

Inferior planets: $\frac{1}{S} = 1 + \frac{1}{P}$

Superior planets: $\frac{1}{S} = 1 - \frac{1}{P}$, where

S is the sidereal period, and
P is the synodic period, both in years.

8-47 The perihelion distance of a planet is its minimum distance from the Sun, and it is given by $a(1 - e)$, where a is the semi-major axis of the orbit and e is the eccentricity.

8-48 The perihelion and aphelion distances from the Sun are given by $a(1 - e)$ and $a(1 + e)$ respectively, where

a is the semi-major axis of the orbit, 0.387 for Mercury, and

e is the eccentricity of the orbit, 0.206 for Mercury.

8-49 (a) Only the superior planets are ever in opposition.

(c) Perihelion distance = $a(1 - e)$

Aphelion distance = $a(1 + e)$

a is the planet's semi-major axis

e is the eccentricity of the orbit.

	Earth	Mars
a	1.0 au	1.52 au
e	0.017	0.093

8-50 See the hint for Exercise 2-20.

SOLUTIONS TO EXERCISES ON THE SOLAR SYSTEM: PLANETS

8-1 (a) Mercury, Venus, Earth, Mars, Jupiter and Saturn.

(b) Venus is brightest at maximum apparent magnitude of -4.3.

(c) The answer varies. It can be any of Venus, Mars, Jupiter or Saturn.

8-2 (a) Photo five; Venus appears largest.

(b) Between 2 and 3.

(c) The absence of features indicates the atmosphere is dense enough to hide surface features, and the ring of light in photo five is caused by atmospheric refraction of sunlight.

8-3 (1) Planets are usually brighter than most stars.

(2) Planets twinkle less than stars.

(3) Planets are located only near the ecliptic--which runs roughly from the east to the west points on the horizon and high through the southern sky--whereas stars are observed in all parts of the sky.

(4) Over a period of several nights, the planets can be observed to move relative to the stars.

(5) If one learns the bright stars in the sky or the patterns of the zodiacal constellations, then the planets can be recognized as extra, temporary bright objects in the sky.

(6) Over a period of several weeks the planets change their brightness; whereas most stars shine with the same brightness every night.

8-4 The orbits of all the planets are near the plane of the ecliptic, and so they all appear to move near that plane (the ecliptic) in the sky.

8-5 It could not be one of the known planets because they never move in the region of the Big Dipper. The Big Dipper is too far north of the ecliptic where the planets are always seen.

8-6 (a) The inferior planets (Mercury and Venus) go through all the phases. The superior planets appear only in the gibbous and full phases.

(b) Mercury is in quarter phase when at greatest elongation.

(c) Jupiter is full at opposition.

8-7 (1) Distance of the planet from the Sun. The closer to the Sun, the brighter the planet. Mercury is closest to the Sun.

(2) Distance of the planet from the Earth. The closer to Earth, the brighter the planet. Venus comes closest to Earth.

8-7 (3) The size of the planet. The larger the planet, the more sunlight it
          reflects. Jupiter is the largest planet.

     (4) The albedo of the planet. The larger the albedo, the greater the per-
          centage of incident sunlight it reflects. Venus has the highest
          albedo, 0.76.

     (5) The phase of the planet. The closer the phase is to full, the brighter
          the planet appears. All the planets are at full phase sometime:
          superior planets when they are closest to or farthest from Earth, infer-
          ior planets when they are farthest.

8-8 (a) A planet changes apparent diameter (observed size) and brightness
          because its distance from Earth varies. In addition, the brightness
          changes because the planet's phase changes.

     (b) You would expect Venus and Mars to go through the greatest changes,
          because their distances (and for Venus its phase) undergo the greatest
          percentage changes of any of the planets.

8-9 (a) Inferior planets are best observed either before sunrise above the
          eastern horizon, or after sunset above the western horizon. Since
          the inferior planets are always seen in the general direction of the
          Sun (never more than 48° away), the only time they can be seen against
          a dark sky is when the E or W horizon lies between the planet in the
          sky and the Sun below the horizon. For example:

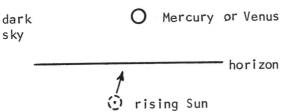

     (b) Outer planets are best observed at night when they are near the
          meridian. That is when they are highest in the night sky, and,
          therefore, that is when the atmospheric distortions are minimized.

8-10 Mercury and Venus can never be observed at opposition, nor can the Earth.

8-11 (a) Sunset   (b) Midnight   (c) Sunrise

8-12 The orbits of the planets do not all lie in the same plane. They are
      inclined enough that the planets often pass above or below the Sun as
      they move through conjunction. Venus at superior conjunction, for example,
      can pass more than a degree above or below the Sun's limb.

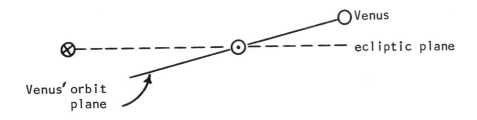

8-13 (a, b) Greatest elongation: inferior planets (Mercury, Venus).

Opposition: superior planets (Mars, Jupiter, Saturn, Uranus, Neptune, Pluto).

8-14 If Jupiter moves 360 degrees in 11.87 years, then in one year Jupiter moves

$$\frac{360^o}{11.87 \text{ yr}} = 30.3 \text{ degrees per year.}$$

Note: Since there are 12 zodiacal constellations around the ecliptic, each occupies about 30°. Thus Jupiter moves through one zodiacal constellation each year. In 1975 it was in Pisces, in 1976 it was observed in Aries, etc.

8-15 (a) Retrograde motion occurs when a faster planet overtakes and passes a slower planet. The slower planet appears to move backward (westward) for a short time against the background stars.

(b) Inferior planets retrograde near inferior conjunction.
Superior planets retrograde near opposition.

8-16 Mercury and Venus are east of the Sun when they are evening stars.

west of Sun

⊙

east of Sun

────⊗────evening horizon

◯ planet

8-17 If the Earth rotates one degree every 4 minutes, then it will rotate 46 degrees in 46 x 4 = 184 minutes.

At western elongation Venus rises before the Sun, so Venus will rise 184 minutes before the Sun at

$$7:07 - 3^h\ 4^m = 4:03 \text{ AM approximately}$$

8-18 (a) 1. Mercury is best observed when it is farthest from the Sun, when it is at greatest elongation.

2. Mercury is also best observed when it is highest above the horizon. This occurs when the ecliptic is nearly perpendicular to the observer's horizon at the time of observation, before sunrise or after sunset.

(b) So the very best time is when conditions (1) and (2) occur simultaneously. The figures below show that in the spring the sunset horizon is nearly perpendicular to the ecliptic (for northern hemisphere observers), and in the fall the ecliptic is nearly perpendicular to the sunrise horizon.

In summary then, Mercury is best observed when it is at greatest eastern elongation in the spring (evening observation) or at greatest western elongation in the fall (morning observation). (For greater accuracy, the eccentricity of Mercury's orbit must be included. Exercise 8-47.)

8-18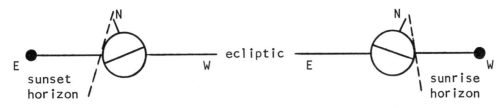

Spring
Mercury at E elongation

Fall
Mercury at W elongation

8-19 Mercury   Closest to Sun.
                 Least known atmosphere.
                 Only planet known to have rotation partially synchronous with
                   revolution.
                 Greatest known density.
                 Most cratered.

      Venus   Hottest surface.
                 Most circular orbit.
                 Comes closest to Earth.
                 Densest CO_2 atmosphere.
                 Highest albedo.

      Earth   Largest satellite relative to planet.
                 Only known planet with liquid water or life.

      Mars   Largest known volcano and longest known canyon.

      Jupiter   Emits more energy than it receives (as does Saturn).
                 Largest in size, mass and number of known natural satellites.
                 Equator closest to its orbit plane.

      Saturn   Extensive ring system.
                 Lowest density.
                 Greatest oblateness.

      Uranus   Greatest inclination of equator to orbit plane.
                 Accidental discovery.
                 Orbit is closest to ecliptic.

      Neptune   Largest perihelion distance.
                 Immediate discovery after existence predicted.

      Pluto   Farthest from Sun.
                 Orbit has greatest inclination and eccentricity.

8-20 There is no water to drink.
      There is no solid surface to stand on.
      There is no Oxygen to breath.
      It is too cold.

8-21 The day temperatures are too hot.
      The night temperatures are too cold.
      There is no atmosphere to breathe.
      There is no water to drink.
      There is no atmosphere to protect from the lethal parts of the solar radiation.

8-22 Mars seems the most likely planet to possess extra-terrestrial life in some
      form. It has a solid surface, some atmosphere, a reasonable temperature
      range, and shows evidence of a history of running water.

8-23 (a) Occasionally Saturn moves in front of a bright star, and the star can
           be observed to shine through the rings.

           Analysis of the spectra of the inner and outer portions of the rings
           shows that the inner moves faster than the outer. Hence the ring
           moves according to Kepler's laws for small orbiting particles, not
           as a solid body.

           Recent radar studies have indicated that the rings are made of solid
           particles on the order of a meter in diameter.

      (b) They are the remnants of a former satellite that got so close to
           Saturn (inside the Roche limit) that Saturn's gravity pulled the
           satellite apart.

           They are material that just never condensed to form one satellite.

8-24 (a) Mars, Jupiter, Saturn.

      (b) The planet's oblateness, and the banded structure of its atmosphere.

8-25 (1) Mars' two polar ice caps change in size during the seasons. The north
           cap disappears entirely when it is summer in that hemisphere, and it
           grows to prominence during winter. The southern cap shrinks in summer
           and grows in winter, but never disappears.

      (2) Each hemisphere changes color during the seasons, appearing lighter
           in the fall and winter and darker in the spring and summer.

8-26 (a) Lunar craters are rugged; Martian craters are more smooth.

      (b) The Martian craters have been smoothed by wind erosion.

      (c) It does have an atmosphere.

8-27 (1) Gravity: the stronger a planet's surface gravity, the greater its
           ability to retain an atmosphere.

      (2) Temperature of the atmosphere: the hotter the molecules in the atmos-
           phere, the faster they are moving, and the more likely they are to
           escape the planet's gravitational pull.

      (3) The mass of the atmospheric molecules: the more massive they are, the
           more energy they require to escape.

8-28 (1) Distance from the Sun: the closer the planet is to the Sun, the more
           energy it receives.

      (2) The rotation period of the planet: the slower the rotation, the more
           time for a given location on the planet to heat up or cool down.

      (3) The ability of the atmosphere to either shield the planet from the
           solar radiation by reflection, or the ability of the atmosphere to
           retain the energy by the greenhouse effect.

8-28 (4) The circulation of the atmosphere: the faster the atmosphere circulates, the more evenly the heat is distributed and the smaller the temperature range.

8-29 We would expect to find fewer craters on Venus for two reasons:

(1) Venus' atmosphere would have destroyed many of the meteorites that would have impacted on the planet.

(2) Venus' atmosphere causes erosion that would destroy craters on the surface.

8-30 On Earth: $F_E = G \dfrac{m_E m_{you}}{r_E^2}$, On Saturn: $F_S = G \dfrac{m_S m_{you}}{r_S^2}$

$$\frac{F_S}{F_E} = \frac{G \dfrac{m_S m_{you}}{r_S^2}}{G \dfrac{m_E m_{you}}{r_E^2}} = \frac{m_S}{m_E} \frac{r_E^2}{r_S^2} = \frac{95}{1} \frac{1}{9.13^2} = 1.1$$

Hence the gravitational attraction on the surface of Saturn is 1.1 times greater than on Earth. If you weigh 150 lbs on Earth, you would weigh 1.1 x 150 = 165 lbs on Saturn.

8-31 (a) The large Martian volcanoes suggest water because it is believed that Earth's water came from volcanoes. Also, Mariner 9 found many features which look like dried up stream and river beds.

(b) Mars has insufficient atmospheric pressure to retain liquid water.

8-32 (a) Inferior planets are those that lie inside the Earth's orbit: Mercury and Venus.

Superior planets are those that lie outside the Earth's orbit: Mars, Jupiter, Saturn, Uranus, Neptune and Pluto.

(b) Terrestrial planets are those with characteristics like the Earth: small, dense, rocky surfaces. Mercury, Venus, Earth, and Mars are terrestrial planets.

Jovian planets are those with characteristics like Jupiter: large, low density, gaseous. Jupiter, Saturn, Uranus and Neptune are jovian planets.

Pluto is probably a terrestrial planet, although its size and density are not well known.

8-33 Stars are hot bodies that produce their own energy; whereas planets are basically cold bodies that reflect energy but produce very little of it.

8-34 (1) The planets are all round and they all rotate.

8-34 (2) The planets are either large and gaseous or small and dense.

    (3) The orbits are nearly coplanar and nearly circular.

    (4) The planets have most of the angular momentum in the solar system.

    (5) Most planets have satellites, with orbits lying near their equatorial planes.

8-35 (a) Oblateness $= \dfrac{48.95 - 45.69}{48.95} = 0.067$

    (b) Distance $= 4.023$ au $\times 1.496 \times 10^8 \dfrac{km}{au} = 6.018 \times 10^8$ km.

    Angular diameter $= 48.95 \times \dfrac{1\ radian}{206265''} = 23.73 \times 10^{-5}$ rad.

    Linear diameter $=$ distance $\times$ angular diameter (radians)

    $= 6.018 \times 10^8$ km $\times 23.73 \times 10^{-5}$ rad.

    $= 142,800$ km.

8-36 Period $= 1.769$ days $\times \dfrac{1\ year}{365.25\ days} = 0.004843$ years

    $a = 421,600$ km $\times \dfrac{1\ au}{1.5 \times 10^8\ km} = 0.002811$ au

    $m_J = \dfrac{a^3}{p^2} = \dfrac{(0.002811)^3}{(0.004843)^2} = 0.947 \times 10^{-3}\ m_{sun}$

    which means that Jupiter is slightly less than $\dfrac{1}{1000}$ the mass of the Sun.

8-37 Uranus was discovered by accident, when an astronomer noted a starlike object that was not on previous star charts.

8-38 Neptune's position in the sky was predicted, based upon the mathematical analysis of the motions of Uranus. Neptune's gravitational attractions caused Uranus to deviate from the path that had been predicted for it without the influence of Neptune. The deviations were used to predict the location of the unknown attracting planet.

8-39 Pluto's location was also predicted from small deviations in the motions of Uranus. Pluto's influence on Uranus is so small, however, that the predications were not accurate, and the search for Pluto required 30 years before the planet was found.

8-40 (a) It was discovered as a streak on a photograph taken during Kowal's search for planets. (See Exercise 8-42).

    (b) Its size is comparable to that of a large asteroid, whereas its orbit is unlike the orbit of any other asteroid.

8-41 It has moved from left of center (in the left photo) to nearly center (in the right photo).

8-42　Clyde Tombaugh, who discovered Pluto, continued to search for additional planets for 12 years, but he found none.

In the early 1970's Brady analyzed the motions of Halley's comet, and found that a large trans-plutonian planet could account for much of the comet's non-gravitational motions. Subsequent searches failed to reveal any planet in the location predicted by Brady.

In 1977, Charles Kowal began a ten-year, systematic search for planets in the zodiacal region of the sky, using the large Schmidt telescope at the Hale Observatory.

8-43　(1)　The planets are remnants of the Sun's exploded binary companion.

Support: a large percentage of the stars in the sky do have binary companions, but none has been found for the Sun.

Contradict: Exploded stars leave a pulsar as a central remnant, but no pulsars have been found in the solar system.

(2)　The planets are the remnants of the collision between the Sun's binary companion and a third star.

Support: same as support for idea number 1.

Contradict: Stellar collisions are very unlikely events.

(3)　The planets were interstellar objects that were captured by the Sun.

Support: Small interstellar dust particles exist.

Contradict: capture is very unlikely, especially of nine objects all in the same plane (ecliptic) moving in the same direction around the Sun.

(4)　The planets formed from material pulled off the Sun during an ancient close encounter with another star.

Support: this would explain the coplanar orbits and motions in the same direction around the Sun.

Contradict: The planets have too much angular momentum.

(5)　The planets formed out of the material left behind when the proto star collapsed to become the Sun.

Support: the planets move around the Sun in the same direction as the solar rotation.

Contradict: Although this currently is the most popular of several questionable hypotheses, it is difficult to understand how the nebular material left behind condensed into planets instead of being dispersed into the galaxy.

8-44 (1) The image of Venus shows no oblateness; whereas the fast rotators are very oblate.

(2) The clouds of Venus show no structure in visible light; whereas the clouds of Jupiter and Saturn show bands of different weather zones that are caused by their rapid rotation.

8-45 (1) The rotation rates of Mercury and Venus were recently determined by analysis of radar signals bounced off the planets.

(2) The rotation speeds of Mars, Jupiter and the Sun were initially determined by noting how long it took specific features to make one rotation around the body.

(3) The rotation rates of Uranus and Neptune were determined from the doppler shift in the light coming from the limbs of the planets.

(4) Pluto's rotation period was recently estimated from periodic variations in the brightness of the distant planet.

8-46 (a) The synodic period of a planet is one revolution relative to the Earth-Sun line--the length of time for the planet to move from one opposition to the next opposition. The sidereal period of a planet is one revolution relative to some point on the celestial sphere.

(b) For a superior planet, $\frac{1}{S} = 1 - \frac{1}{P}$, where

S is the sidereal period and
P is the synodic period.

$$\frac{1}{12} = 1 - \frac{1}{P} \quad \text{or} \quad \frac{1}{P} = 1 - \frac{1}{12} = \frac{12 - 1}{12} = \frac{11}{12}$$

$P = \frac{12}{11} = 1\frac{1}{11}$ years, for Jupiter.

(c) For inferior planets, $\frac{1}{S} = 1 + \frac{1}{P}$

$$\frac{1}{S} = 1 + \frac{1}{1.6} = \frac{1.6}{1.6} + \frac{1}{1.6} = \frac{2.6}{1.6}$$

$S = \frac{1.6}{2.6} = 0.615$ years $= 225$ days, for Venus.

8-47 The planet is obviously Neptune or Pluto; both perihelion distances must be calculated to determine which:

	Neptune	Pluto
a	30.2 au	39.4 au
e	0.01	0.25
a(1 - e) =	30.2 x 0.99	39.4 x 0.75
	= 29.9 au	= 29.6 au
	= 44.9 x 10^8 km.	

Therefore Neptune has the largest perihelion distance, 29.9 au.

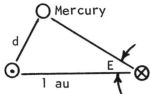

8-48

Mercury

d

1 au

E

The angle E is the elongation of Mercury. It is greatest when Mercury is simultaneously at aphelion and at maximum elongation. Then

$$d_{max} = a(1 + e) = 0.387(1 + 0.206) = 0.467 \text{ au.}$$

$$E_{max} = \sin^{-1}\left(\frac{d}{1\text{au}}\right) = \sin^{-1} 0.467 = 28°$$

The angle E is minimum when Mercury is simultaneously at greatest elongation and at perihelion. Then

$$d_{min} = a(1 - e) = 0.307 \text{ au.}$$

$$E_{min} = \sin^{-1} 0.307 = 18°.$$

8-49 (a) Mars comes closest to Earth in opposition.

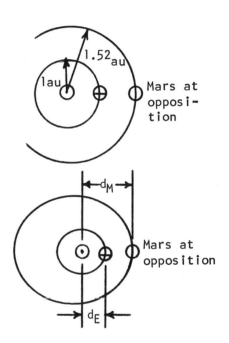

(b) Assuming circular planetary orbits, the minimum Earth-Mars distance is just $a_{Mars} - a_{Earth}$ = 1.52 au - 1 au = 0.52 au.

(c) The closest the planets can ever approach each other is when the Earth is at aphelion in its orbit and Mars is at perihelion.

At aphelion, the Earth-Sun distance is

$$d_E = a_E(1 + e_E)$$
$$= 1.0(1 + 0.017) = 1.017 \text{ au.}$$

At perihelion, the Mars-Sun distance is

$$d_M = a_M(1 - e_M) = 1.52(1 - 0.093) = 1.379 \text{ au.}$$

Thus the minimum separation is $d_M - d_E = 0.362 \text{ au.}$

8-50 For Uranus, $P = 83.74$ years

$$a = 19.14 \text{ au.}$$
$$P^2 = (83.74)^2 = 7012$$
$$a^3 = (19.14)^3 = 7012 \qquad \text{check}$$

Asteroids

9-1 Where are most of the asteroids located in the solar system?

9-2 (a) Describe the shapes of typical large and small asteroids.

(b) How is the shape of an asteroid estimated?

9-3 Cite some reasons that astronomers study asteroids, and mention some uses to which their studies have been put.

9-4 Cite several hypotheses that explain why there are many small asteroids instead of a single planet in the region between Jupiter and Mars. For each hypothesis cite one supporting and one contradicting argument or idea.

9-5 Suppose an asteroid has an aphelion distance = 3 au.
perihelion distance = 1 au.

(a) Sketch the orbits of the Earth, Mars, Jupiter and the asteroid.

(b) Calculate the semi-major axis of the asteroid orbit (in au's).

(c) Calculate the period of the asteroid in years.

(d) On the asteroid which would be the longest season, in terms of human comfort, summer or winter?

9-6 (a) In what way might an unmanned mission to an asteroid (or comet) be easier to perform than a similar mission to one of the planets?

(b) In what way might it be more difficult to perform?

9-7 Cite some experiments that would likely be performed by an unmanned space-craft on a mission to an asteroid.

9-8 Would you expect to find evidence of liquid water or of an atmosphere on an asteroid? Explain your answer.

Comets

9-9 Cite two ways that comets are often discovered.

9-10 Where are most of the periodic comets located in the solar system?

9-11 (a) What is the mass of a typical comet?

      (b) Why are the masses of comets so difficult to measure?

9-12 (a) What elements are commonly detected in comets?

      (b) How are those elements detected in the comets?

9-13 (a) What is the probable structure of a comet when it is outside the orbit of Jupiter?

      (b) What is the structure of a comet when it is inside the orbit of the Earth?

9-14

Exercise 9-14.

An object photographed within the solar system. (Lick Observatory photo.)

The photo is of an object found within the solar system.

      (a) What is the object?

      (b) Explain the diagonal white streaks in the photo.

9-15 How is it known that comets are composed of small solid particles in addition to gases? Two reasons.

9-16 (a) What would be the effect (on the Earth) of the Earth passing through the tail of a comet?

(b) What would be the effect of the Earth colliding with a comet nucleus?

9-17 (a) What is a possible relationship between comets and meteors?

(b) What is a possible relationship between comets and asteroids?

9-18 Cite some ways that the Sun influences a typical comet.

9-19 Why do comet tails generally point away from the Sun?

9-20 (a) From what location (in or around the solar system) do comets come?

(b) From what source did cometary material probably originate?

9-21 Why do periodic comets tend to have relatively short life spans?

9-22 Which planet seems to be responsible for "capturing" the most comets, and why this one?

9-23 (a) How might the orbit of a comet be substantially changed?

(b) Is there any evidence that this has actually happened?

Meteors

9-24 Distinguish between a meteor, meteorite and meteoroid.

9-25 What is the glow or the light that is called a meteor?

9-26 (a) What is the cause of sporadic meteors?

(b) What is the cause of shower meteors?

9-27 Are you more likely to observe more sporadic meteors in the AM hours or in the PM hours? Why?

9-28 How could one detect a meteor shower that occurs in the daytime?

9-29 What is wrong with the term "meteor crater"?

9-30 Meteor camera A is located 100 km due south of meteor camera B, when a meteor is simultaneously photographed by both cameras. According to camera A, the meteor began at an altitude of 45° in the due north direction (azimuth 0°); whereas camera B saw the meteor begin at the zenith. Where did the meteor begin?

Meteorites

9-31 Why do scientists study meteorites?

9-32 What kind of meteorite are you most likely to recognize during a search in the desert, and why?

9-33 Although stony meteorites are the most common to fall, why are they hard to discover?

9-34 It is estimated that each day the Earth accretes several thousand tons of material from space. Why, then, don't we all have to wear hard hats to avoid being struck when we're outside?

9-35 (a) Which members of the solar system (planets and satellites) are known to have craters on them that might have been caused by meteorite impacts?

(b) Comment on whether or not you would expect those planets not cited in part (a) to have impact craters on their surface.

9-36

Exercise 9-36.

Iron Meteorite (Photo by Griffith Observatory)

What is it about the appearance of the rock in the photo which clearly identifies the rock as a meteorite?

Meteoroids

9-37 (a) What is the physical difference between meteoroids that produce fireballs and those that produce sporadic meteors?

(b) Which is more likely to be found (as a meteorite) on Earth?

9-38 Why are the meteoroids which produce fireballs thought to be associated with asteroids; whereas those which produce sporadic meteors are not associated specifically with asteroids?

9-39 How is it known that the density (number per cubic meter) of small meteoroids is approximately constant throughout the inner solar system (within Jupiter's orbit)?

HINTS TO EXERCISES ON ASTEROIDS, COMETS AND METEORITES

9-1 Most are in what is called the "asteroid belt."

9-2 Consider how its shape might influence the amount of light the asteroid reflects.

9-3 One use involves the masses of the planets that influence the asteroid's motion.

9-4 Some hypotheses suggest that they are remnants of other objects commonly found in the solar system.

9-5 Review elliptic motion, Exercises 2-19 and 2-20.

9-6 Some asteroids pass close to the Earth, but they exert little gravitational attraction.

9-7 The lack of surface gravity on the asteroid will make it very difficult to land on the asteroid.

9-8 Consider what environment is needed to retain water and an atmosphere on an orbiting body. See Exercise 8-27.

9-9 One involves astronomical photography, the other involves visual searches with small telescopes.

9-10 Their orbits tend to lie near the ecliptic plane.

9-11 The mass of an orbiting body is usually determined from its gravitational influence on another orbiting body.

9-12 Elements are usually determined from analysis of spectra.

9-13 Outside the orbit of Jupiter the comet gets very little solar radiation.

9-14 The photo is a time exposure of an object which changes its appearance as it nears the Sun.

9-15 Consider shower meteors, and the spectra of comets far from the Sun.

9-16 The nucleus of a comet is probably made of small particles and frozen gases.

9-17 (a) Consider shower meteors.

9-18 The influences involve the Sun's gravity, radiation and the solar wind.

9-19 Consider the effect of the solar wind on the coma.

9-20 Comets are believed to enter the solar system on orbits that are hyperbolic or very long ellipses.

9-21 Consider what happens to a comet when it nears the Sun.

9-22 A planet's gravitational attraction depends upon its mass.

9-23 Consider the effect of a close encounter with some large object.

9-24 Two are pieces of matter; the third is a flash of light.

9-25 It is not a glowing rock.

9-26 One involves comet remnants; the other random meteoroids.

9-27 Consider the locations of the AM and PM observers on the revolving and rotating Earth.

9-28 The ionized air along the meteor trails will reflect radio waves.

9-29 Review the differences between meteors and meteorites.

9-30

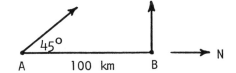

9-31 Consider what is unique about them.

9-32 Meteorites are usually found because of some special property that allows
9-33 the searcher to distinguish them from ordinary rocks.

9-34 Much of the accreted material is in the form of micrometeorites.

9-35 (b) Remember that a planet must have a hard surface in order to retain
        impact craters.

9-36 It is the Widmanstatten figures.

9-37 Consider their sizes.

9-38 Consider the orbits of the meteoroids before they reach Earth.

9-39 The evidence has been gathered by interplanetary spacecraft.

SOLUTIONS TO EXERCISES ON ASTEROIDS, COMETS AND METEORITES

9-1   Between the orbits of Mars and Jupiter.

9-2   (a)   The larger asteroids (like Ceres) are believed to be spherical, while
             the smaller ones are of irregular shapes.

       (b)   From studies in the variations in the light reflected from the tumbling
             asteroids.

9-3   They are there and astronomers are curious about them.
       They may provide clues regarding the origin of the solar system.
       They have served as probes into the gravitational fields of the solar system,
          to provide information on planetary masses.
       They have been used to improve the fundamental celestial coordinate system.

9-4   Remnants of an exploded planet:
       For: The orbits are like planetary orbits (1).
       Against: The total mass of all the asteroids is much less than the mass of
                  a small planet (2), and there is no known reason why a planet
                  should explode.

       Remnants of collided planets:
       For: Reason (1).
       Against: Collision is a very unlikely event, and reason (2).

       "Dead" comets
       For: A few have orbits like comets (e.g., Hidalgo, Icarus), and the smaller
             asteroids may be about the size of a comet nucleus.
       Against: Most do not have cometary orbits, and there's no reason why so
                 many dead comets would collect in one region.

9-5   (a)

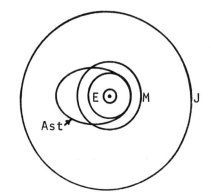

       (b) $a = \dfrac{3 \text{ au} + 1 \text{ au}}{2} = 2$ au.

       (c) $P^2 = a^3 = 2^3 = 8$

             $P = \sqrt{8} = 2.8$ years

       (d)   Winter, since the asteroid spends
             most of its time outside the
             orbit of Mars (Kepler's second
             law).

9-6   (a)   It would be easier because several asteroids and comets come much
             closer to Earth than do any of the planets.

       (b)   It would be more difficult because the objects are relatively small
             and exert very little gravitational attraction.

9-7   The spacecraft would photograph the asteroid, and measure its magnetic field, gravitational attraction, size, and emitted and reflected radiations. If possible a sample would be analyzed for structure and composition.

9-8   Neither would be expected, because there is insufficient gravity to retain either liquids or gases on or around any of the asteroids.

9-9   They are discovered as fuzzy images on photographs, taken usually for some astronomical research other than comets.

      They are also discovered by amateur astronomers who use relatively small telescopes to regularly search the skies for new or unusual objects.

9-10 Most have elongated elliptical orbits that carry them from the region of the terrestrial planets to the region of the large planets.

9-11 (a) They are estimated to range from 10^{10} to 10^{12} tons.

      (b) Cometary masses are too small to measure directly, since no comet has ever been observed to exert a measurable gravitational influence on another body.

9-12 (a, b) Bright emission lines of H, O, C and N are usually found in cometary spectra when they get within $\sim$3 au of the Sun.

9-13 (a) The popular model (which is not universally accepted) is that of frozen gases, dust and rocks.

      (b) A small nucleus of frozen gases, dust and rocks; surrounded by a coma of gas and dust that has evaporated from the nucleus. A tail of gas and dust that extends for millions of kilometers away from the nucleus and coma. Recent spacecraft observations have found a large cloud of hydrogen surrounding the comas of comets.

9-14 The object in the center of the photo is a comet (Humason, 1961a), and the streaks are stars.

9-15 When a comet is at a large distance from the Sun (greater than 5 au) its spectra usually resembles that of the Sun, which indicates that it is made of solid particles that reflect sunlight. Meteor showers are believed to occur when the Earth encounters small solid comet remnants.

9-16 (a) At most an increase in the number of meteors observed for a day or two.

      (b) Collision would probably cause a large crater if it struck dry land.

9-17 (a) It is thought that meteor showers result when the Earth encounters the remnants of a comet.

      (b) Some people have suggested that a few of the asteroids (those with very eccentric orbits) might be "dead" comets, but this idea is not widely accepted.

9-18 The Sun's gravity influences the comet's orbit.
      Its radiation heats up the nucleus causing gases to evaporate.
      The radiation pressure and solar wind help create the tail.

9-19 The solar wind and radiation pressure act on the evaporated gases and dust of the coma to gently push the particles radially outward from the Sun and away from the comet's head.

9-20 (a, b) The answers are not known for certain. The most popular hypothesis is that of the comet cloud. This would be a shell of interplanetary material that was left over after the planets were formed, and which has been pushed out to 50,000 au (approx.) from the Sun. Occasional passing stars perturb this material so that it falls in toward the Sun.

9-21 Each time a comet passes perihelion part of its mass is evaporated into space, and after several passages it is evaporated away.

9-22 Jupiter is primarily responsible, since it is the most massive planet.

9-23 (a, b) It is believed that comets first enter the solar system on hyperbolic or very long elliptic orbits. A few pass close enough to one of the major planets to enable the planet's gravity to change the comet orbit into a small ellipse. The periodic comets, such as Halley's and Encke's are examples.

9-24 Meteoroid: Small particle (rock) in space.

Meteorite: A meteoroid that has landed on a planet or a moon.

Meteor: The flash of light caused by a meteoroid travelling through an atmosphere.

9-25 As the small meteoroid (often no larger than a sand grain) passes through the Earth's upper atmosphere at very high speeds, it ionizes the atoms in the air, causing the air to glow.

9-26 (a) The Earth encountering the random debris in the solar system.

(b) The Earth encountering the remnants of a comet as it passes through the orbit of a comet.

9-27 You are likely to observe more in the dark AM hours, because then you are on the Earth's leading side as it moves around the Sun.

9-28 They can be detected by radar signals which reflect off the ionization trails of the meteors.

9-29 Since a meteor is a flash of light, it can hardly produce a crater. The proper term is "meteorite crater."

9-30 It began 100 km above camera B.

9-31 Except for Moon rocks, they are man's only extraterrestrial samples.

9-32 Iron meteorites are more easily discovered because they (a) tend not to break up into small pieces, (b) are unusually dense, and (c) can be detected with a metal detector.

9-33 They often break up into small fragments, and they are hard to distinguish from ordinary terrestrial rocks.

9-34 Most of the accreted material is in the form of micrometeorites--small dust-sized particles which settle to the ground too slowly to be recognized. Also, most of the larger meteor-producing meteoroids burn up in the atmosphere before reaching the ground.

9-35 (a) Mariner spacecraft have found craters on Mercury, Mars and the two satellites Phobos and Deimos. Radar studies indicate the presence of craters on the surface of Venus, and craters have been seen on the Earth and the Moon.

(b) We don't expect to find craters on the jovian planets since they probably do not have hard surfaces. Pluto may have some.

9-36 The crossed diagonal (Widmanstatten) lines appear only in meteorites.

9-37 (a) Most meteors are caused by very small grains; whereas fireballs are caused by larger meteoroids.

(b) The large meteoroids are more likely to survive the trip through the atmosphere. Most of the small grains burn up.

9-38 The large fireball-producing meteoroids generally have orbits that are similar to asteroid orbits--low inclination to the ecliptic and low eccentricity--whereas the orbits of the smaller particles seem to have no preference for the ecliptic plane.

9-39 The Mariner, Pioneer, Viking and Voyager spacecraft carry meteorid detectors, and they have found this result. In particular the Pioneers found the surprising result that there was no increase in the number of small particles encountered as they flew through the asteroid belt on the way to Jupiter.

General Properties

10-1   What are the overall properties of the Sun: size, mass, surface temperature, central temperature, approximate age, rotation rate and composition?

10-2   What property identifies the Sun as a star rather than a planet?

10-3   Describe two motions of the Sun.

10-4   State two different ways to determine the solar rotation rates, and give the values of the rotation rates.

10-5   Calculate the rotation rate of the Sun, given that the observed (from Earth) radial velocities of its east and west limbs differ by 3.8 km/sec.

10-6   How is the Sun's age estimated?

10-7   In what ways does the study of the Sun contribute to our knowledge of the stars?

10-8   (a) If the surface temperature of the Sun is $6000^{\circ}K$, use Wien's law to estimate the wavelength at which the maximum amount of solar energy is radiated. What color is this radiation?

(b) If the Sun got hotter on the surface, would it become red or white?

*10-9   Calculate the "solar constant" at the orbit of Saturn. Compare your answer to the value at the Earth, 1.4×10^6 erg/cm^2-sec, to estimate how much less solar energy is available for spacecraft flying to the outer planets.

Appearance of the Sun and Its Influence on the Earth

10-10   (a) Why does the Sun's apparent (observed) size vary throughout the year?

(b) During what month does the Sun appear largest?

10-11   What are several ways that the Sun influences the Earth?

10-12   Why have Earth creatures evolved with organs that are sensitive to visible light? Why not to radio or ultraviolet, for example?

10-13   What are two ways that solar flares influence the Earth?

10-14   What was the advantage of observing the Sun from Skylab?

Surface Features: Spots and Assorted Blemishes

10-15   (a)   What are sunspots, and how long do they last on the average?

        (b)   What can be determined from the motions of sunspots?

10-16   If the Sun's image projected on a screen is 25 cm in diameter, and the diameter of a sunspot on the image is 3 mm (0.3 cm), what is the actual diameter of the sunspot in kilometers?

10-17   The Sun rotates in a counterclockwise direction, once about every 27 days. Do we observe sunspots to move across the face of the Sun from east to west or from west to east?

10-18   Herschel suggested that sunspots are holes in the Sun's fiery envelope, through which we could observe the inhabited rocky interior. Cite evidence to refute Herschel's idea.

10-19   What would be the appearance of the Sun in the sky if it were completely covered by a giant sunspot?

10-20   During the 11-year sunspot cycle, in what ways do each of the following change or vary?

        (a)   The number of spots,
        (b)   The locations of the spots,
        (c)   The sizes of the spots.

10-21   (a)   From the diagram, estimate the average duration of the sunspot cycle during the period 1910 to 1970.

        (b)   During what year do you expect the next sunspot maximum?

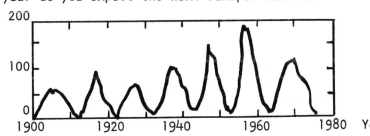

10-22   (a)   What observation indicates the presence of magnetic fields in sunspots?

        (b)   What is the relationship between magnetic fields and spots?

10-23   Why is it said that the true solar cycle is 22 years rather than the 11-year sunspot cycle?

10-24   Suppose that an elastic freeway were attached to the surface of the Sun (assuming that were possible), running directly along a meridian from one pole of the Sun to the other. Sketch and explain the appearance of the road one month after it was built.

10-25 What is the photosphere of the Sun?

10-26 (a) What are the granulations on the Sun?

      (b) What evidence supports your answer to part (a)?

10-27 Cite two pieces of evidence that the Sun is neither a solid nor a liquid body.

10-28 Why does the Sun appear darker around its limbs than in its central region?

10-29 (a) Where on the Sun are prominences observed?

      (b) Under what conditions can prominences be observed without instruments.

      (c) What role do magnetic fields play in the appearance of prominences?

10-30 (a) From a picture of a prominence in your text, estimate the height of the prominence above the photosphere (in kilometers).

      (b) Why is the prominence probably higher than your estimate?

10-31

Exercise 10-31.
The Sun.
(Lick Observatory photo.)

      (a) The photo is of which portion of the Sun.

      (b) During what event is this portion of the Sun visible to the naked eye?

      (c) Why is it not naturally visible to the naked eye at other times?

      (d) What instrument allows the astronomer to continuously study this portion of the Sun?

10-32   In what ways does the Sun eject matter into space?

10-33   It has been suggested that men could be landed on the Sun, provided the landing took place at night. Why can't this (jestful) suggestion ever be carried out?

Energy Production and Transfer--The Solar Interior

10-34   (a)   How is energy produced in the Sun?

         (b)   In what form is the energy which is produced?

         (c)   How is the visible light produced in the Sun?

10-35   Neutrinos produced in the core of the Sun reach the photosphere and escape in less than 3 seconds. Why does it take gamma and X-ray photons on the order of a million years (on the average) to make the same trip from the solar core to the surface?

10-36   (a)   Briefly describe the three energy-transfer processes of conduction, radiation and convection.

         (b)   Cite one example of each process.

         (c)   Which is the slowest process?

10-37   The pressures in the center of the Sun are estimated to be between 10 and 20 billion lbs per square inch. How is it that the material there can be in the gaseous state?

10-38   What is the major problem that scientists face in trying to reproduce the Sun's basic energy process (nuclear fusion) for peaceful use on Earth?

*10-39   The Sun emits 5×10^{23} horsepower of energy. Calculate how many horsepower of solar energy fall on a typical suburban lot (50 x 100 ft) when the Sun is at the zenith? Assume half the available power is lost to atmospheric absorption and reflection.

*10-40   To generate its energy, the Sun converts hydrogen into helium, with a small percentage of the mass converted into energy. Specifically, when one gram of H is converted, 0.993 grams of He are produced, and 0.007 grams are converted into 6.4×10^{18} ergs of energy (enough to raise the Apollo-Saturn V system 12 miles above the ground). During the Sun's 5 billion year life what percent of its total mass has been converted into energy?

*10-41   At the present rate of energy production, how long would it take to convert the entire Sun from hydrogen into helium?

HINTS TO EXERCISES ON THE SUN

10-1   Look these up in the index of the text.

10-2   See Exercise 8-33 for the basic difference between stars & planets.

10-3   Like the Earth, it both rotates and revolves.

10-4   The methods involve Doppler shift and sunspots.

10-5   Exercise 10-1 gives the solar diameter.   There are 86400 sec/day.

10-6   The Sun is believed to be about the same age as the oldest known objects
        in the solar system.

10-7   The Sun is an ordinary star, only much closer than the others.

10-8   Wien's law is given in Exercise 3-26.

10-9   The value varies inversely as the square of the distance to the planet.
        Saturn is 9.5 au from the Sun.

10-10   Consider the shape of the Earth's orbit.

10-11   Its primary influences are by gravitation and radiation.

10-12   Selective evolution would dictate that creatures who detect whatever radia-
        tion is most abundant would have the advantage.

10-13   These involve the aurora and radio communications.

10-14   Skylab orbited the Earth above the atmosphere.

10-15   The average temperature of a spot is 1500 to 2000 degrees cooler than the
        photosphere.

10-16   See Exercise 10-1 for the solar diameter.

10-17   The Earth's rotation and revolution play no role in this question.

10-18   Consider your answers to Exercise 10-1.

10-19   See the hint to Exercise 10-15.

10-20   The answer to part (a) can be found in Exercise 10-21.

10-21   Determine the time intervals between each successive peak, or between each
        successive valley, and take their average.

10-22 Consider the Zeeman effect and the polarity of spot pairs.

10-23 Consider the magnetic fields and polarities of the spots.

10-24 Consider the differential rotation rates of the Sun.

10-25 It is one of the observable layers of the Sun.

10-26 They are caused by the process of convection which transports energy to the surface of the Sun.

10-27 Consider its temperatures, sunspot motions and surface activity.

10-28 The deeper one sees into the Sun, the brighter it appears.

10-29 They are arches of relatively cool (compared to the corona) gas suspended above the photosphere.

10-30 The solar diameter is given in Exercise 10-1.

10-31 (c) This part of the Sun is not normally seen because the brighter photosphere overwhelms its feeble light.

10-32 It does this continually, but more so during its active phase.

10-33 Consider what is meant by night on the solar surface.

10-34 The energy production occurs in the core of the Sun; whereas we observe the energy that escapes from the surface.

10-35 Neutrinos travel through the entire Sun without interacting (by absorption and reemission) with the solar material; whereas photons interact with the solar material.

10-36 All can occur in the process of bringing a pan of tomato soup to a boil on an electric burner.

10-37 Consider the central temperature of the Sun.

10-38 Consider the extreme temperatures and amounts of energy produced by the reactions.

10-39 The answer is on the order of a large American gas guzzler. Assume the Earth is 5×10^{11} feet from the Sun.

10-40 The total solar energy output = 4×10^{33} ergs/sec. One year is about 3×10^{7} sec., and the solar mass = 2×10^{33} grams.

10-41 Solve Exercise 10-40 first to determine the grams of hydrogen per second which are converted to helium + energy.

10-1   Diameter: 864,000 mi.   Surface temp: $6000°K$ (approx.)
                    1,390,000 km.   Central temp: 16 million$°K$ (approx.)
        Mass: 2×10^{33} grams.   Composition: 75% hydrogen
        Age: 4.6 billion yrs. (approx.)   24% helium (approx.)
        Rot. rates: 25 to 35 days.

10-2   The Sun produces its own light and energy; whereas planets mostly reflect light and energy.

10-3   The Sun rotates once every 25 to 35 days, depending on the solar latitude. It orbits the galactic nucleus at a speed of 200 to 300 km/sec, making one revolution about every 200 million years.

10-4   (1)   Observe the Doppler shift in the light coming from the approaching and receding limbs of the Sun.

        (2)   Observe how long it takes for sunspots to move across the face of the Sun.

    The rate varies from 25 days near the Sun's equator up to 35 days near the polar regions.

10-5   Circumference = π × diameter = 4.36×10^6 km.

    Time for one rotation = $\dfrac{4.36 \times 10^6 \text{ km}}{1.9 \text{ km/sec}}$ = 2.3×10^6 sec = 26.6 days.

10-6   It is estimated to be about the age of the oldest known Moon rocks and meteorites.

10-7   The Sun is the only star close enough to examine in detail--the only star upon which we can observe details of spots, flares, granulations, prominences, etc.

10-8   (a) λ_{max} = 0.29/T = 0.29/6000 $°K$ = 0.000048 cm = 4800 $Å$ (yellow).

        (b)   White.

10-9 $\dfrac{c_{Sat}}{c_{\oplus}} = \left(\dfrac{a_{\oplus}}{a_{Sat}}\right)^2$; $c_{Sat} = 1.4 \times 10^6 \times \left(\dfrac{1}{9.5}\right)^2 = 1.55 \times 10^4$ ergs/cm^2-sec.

    Hence there is 90 times less energy per cm^2 per sec at Saturn.

10-10   (a)   Because the Earth-Sun distance varies due to the Earth's elliptic orbit.

        (b)   January when the Earth is closest to the Sun.

10-11   Its gravity keeps the Earth in orbit and influences ocean tides.
        Its energy sustains life and drives the water cycle.
        Its light illuminates our environment.

        Its flares disrupt long-range radio communications.
        Its charged particles contribute to the aurora and the Van Allen radiation belts.

10-12 Most of the solar output is visible light, to which the atmosphere is transparent. Hence there is more visible light naturally available on the Earth's surface than other kinds of radiation.

10-13 (1) High-energy charged particles given off by a flare interact with the upper atmosphere to temporarily disrupt long-range radio communications (which depend upon being bounced off the ionosphere).

(2) The charged particles are deflected by the Earth's magnetic field lines and eventually spiral into the upper atmosphere at high latitudes where they interact with the air molecules to produce the aurora.

10-14 From Skylab observations could be made of all the radiation emitted by the Sun; whereas ground-based observations can be made of only the radiation which passes through the atmosphere.

10-15 (a) They are relatively cool regions of the photosphere which last a few days on the average, although a few last several weeks.

(b) Their observed motions are used to estimate the solar rotation rates.

10-16 $$\frac{0.3 \text{ cm}}{25 \text{ cm}} = \frac{\text{spot diameter}}{\text{Sun's diameter}}$$

spot dia. $= \frac{0.3}{25} \times 1390,000$ km $= 16,700$ km., slightly larger than the Earth.

10-17 East to west.

10-18 With modern instruments we can observe that spots are not holes, and that their temperatures are about 4000 °K, much too hot to sustain life. Also the gravity on the Sun is too strong for inhabitants.

10-19 It would appear as a reddish-orange star, still quite bright.

10-20 (a, b) The yearly number of spots goes from a mimimum of 5 to 10 (occurring near latitudes 30 to 40 degrees), to a maximum of 50 to 100 (occurring near latitudes 15 to 25 degrees), to a minimum of 5 to 10 (occurring on either side of the solar equator).

(c) On the average the spots are over twice as large at sunspot maximum as at minimum.

10-21 (a) Using the peaks:
1917 - 28   11 years
1928 - 38   10
1938 - 47   9
1947 - 58   11
1958 - 69   11
            ‾‾‾‾‾‾
            10.6 ave.

(b) 1969 + 11 = 1980

10-22 (a) The Zeeman effect is quite evident in sunspot spectra.

(b) In a pair the two spots have opposite polarity, leading to the idea that a magnetic field loops into one spot and out the other.

10-23 The magnetic polarities of the leading and trailing spots of all the pairs in a hemisphere reverse every 11 years, so that the period over which the number, location and magnetic field all repeat is 22 years.

10-24 The Sun rotates differentially, faster in the equatorial regions than in the polar regions.

Road after a month.

10-25 It is the surface of the Sun that we see. It is the region where most of the photons of visible light are released into space.

10-26 (a) The light areas are the tops of rising hot columns of gas; the darker regions are cooler gas sinking back into the Sun.

       (b) Doppler shift indicates that the light areas are moving up toward the surface and the darker regions are sinking into the solar interior.

10-27 At the surface temperatures of the Sun, all known substances are gases. If the Sun were a solid, all the spots would appear to move around the Sun at a uniform rate rather than faster near the equator and slower in the polar regions. Also, material on the surface of the Sun is observed to move, which could not happen on a solid body.

10-28 When observing the solar limb one sees regions that are farther from the Sun's core and hence cooler and darker; whereas when observing the center of the Sun's face one sees into deeper, hotter, brighter regions of the photosphere.

10-29 (a) On the limbs of the Sun.
       (b) During a total solar eclipse.
       (c) Magnetic field lines are believed to give the prominences their arch shape and help to keep the material suspended in the corona.

10-30 (a) Typical heights run from about 100,000 to 500,000 km.

       (b) The lower part of the prominence probably extends below the limb in your picture.

10-31 (a) Solar corona.
       (b) Total solar eclipse.
       (c) The bright photosphere overwhelms the corona.
       (d) Coronagraph.

10-32 The solar wind consists of particles from the Sun, continuously pushed out into the solar system by solar radiation pressure. Flares eject large quantities of matter into space.

10-33 There is no night on the Sun.

10-34 (a) Energy is produced in the core of the Sun by the conversion of hydrogen into helium + energy.

       (b) Energy produced in the core is in the form of neutrinos and very energetic electromagnetic radiation: gamma rays and X-rays.

       (c) In making the long journey from the core to the photosphere the gamma and X-rays lose much of their energy, thus becoming less energetic photons such as visible light and infrared.

10-35 In the Sun gamma and X-ray photons travel only about a centimeter (on the average) before they are absorbed by a hydrogen or helium nucleus and then emitted in some random direction. They gradually work their way to the surface by this process of continual absorption and emission. Neutrinos, on the other hand, usually pass out of the Sun without being absorbed by anything.

10-36 (a, b) Within matter energy is transferred by conduction when vibrational molecular energy is transferred to adjacent, less energetic molecules. For example, heat is transferred along a poker left in a fire until eventually the handle gets hot.

Energetic photons move from one region to another by radiation, as when light moves from the Sun to the Earth.

In the convection process, energetic matter is moved from one place to another, as when hot soup on a stove boils up from the bottom of the pan (if it doesn't stick).

(c) Conduction is the least efficient and requires the longest time to transport energy.

10-37 At the central temperatures of the Sun, about 16 billion degrees, everything is in the gaseous state.

10-38 It is difficult to construct a "box" that can withstand the high temperatures of nuclear fusion and contain the energy produced.

10-39 The area of a sphere of radius 1 au is $4\pi r^2 = 4\pi(5 \times 10^{11})^2$
$$= \pi \times 10^{24} \text{ ft}^2.$$

The solar energy crossing each square foot is $\dfrac{5 \times 10^{23} \text{ hp}}{\pi \times 10^{24} \text{ ft}^2} = 0.16 \text{ hp/ft}^2$.

The 50 x 100 lot has 5000 ft^2, so the total energy falling on that lot is

$\frac{1}{2} \times 0.16 \text{ hp/ft}^2 \times 5000 \text{ ft}^2 = 400 \text{ hp}$.

10-40 If each gram of H produces 6.4×10^{18} ergs, then in one second

$\dfrac{4 \times 10^{33} \text{ ergs/sec}}{6.4 \times 10^{18} \text{ ergs/gm}} = 6.25 \times 10^{14}$ gm/sec of H is converted to He and energy.

Of this, 6.25×10^{14} gm/sec x 0.007 = 4.4×10^{12} gm/sec is converted to energy.

In 4.6 billion years there are 1.4×10^{17} secs, so that during this time 4.4×10^{12} gm/sec x 1.4×10^{17} sec = 6.2×10^{29} gm is converted to energy. This is 0.031% of the total solar mass.

10-41 $\dfrac{2 \times 10^{33} \text{ gm}}{6.25 \times 10^{14} \text{ gm/sec}} = 3.2 \times 10^{18}$ sec $\approx$ 100 billion years.

Stellar Parallax and Distances

11-1   If a star's parallax is 0".04, what is its distance in parsecs, in light years, in astronomical units, in kilometers?

11-2   (a)   How is the parallax of a star measured?

   (b)   What is the largest known stellar parallax?

   (c)   Calculate the distance to the star mentioned in part (b).

11-3   The smallest parallaxes that can be measured are about 0".01. What is the maximum distance that can be measured by the method of stellar parallax?

11-4   What would be the advantage of measuring stellar parallaxes from Mars rather than from Earth?

11-5   (a)   Calculate the distance to the star whose parallactic motions are shown in the diagram.

   (b)   The fact that the parallactic motion is an ellipse, and not a straight line or a circle, tells what about the star's position in the sky?

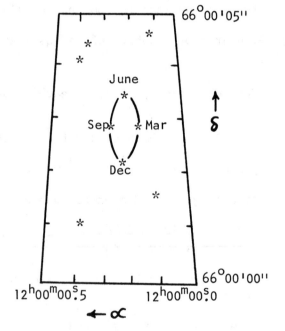

11-6   If the Sun were 3 cm. in diameter-- about the size of a ping pong ball-- how far away from the Sun, on this scale, would be

   (a)   the Earth,
   (b)   Pluto,
   (c)   the nearest star,
   (d)   the great galaxy in Andromeda?

Observed Stellar Motions

11-7   (a)   How is the proper motion of a star determined?

   (b)   What is the largest known proper motion for a star?

11-7   (c)   What is the rough correlation between a star's proper motion and its distance?

11-8   About how long will it take for the bright star Procyon to move $\frac{1}{2}$ degree across the sky, a distance equal to the angular diameters of the Sun and Moon.

*11-9   If a star in the Andromeda galaxy has a transverse space motion of 200 km/sec , how long would we have to observe the star before its proper motion could be detected? Assume the smallest proper motion that can be detected is 0''.01.

11-10   Why can't we determine the proper motions of the stars in the great galaxy in the constellation of Andromeda?

*11-11   If a star is 40 parsecs away and has a proper motion of 0''.05 per year, what is the star's speed (in km/sec) across the observer's line of sight?

11-12   (a)   How is the radial velocity of a star determined?

        (b)   What is the average radial velocity of the stars in the solar neighborhood?

11-13   Suppose the red hydrogen line Hα in the spectrum of a star appears at $\lambda = 6565$ Å .

        (a)   What is the star's radial velocity?

        (b)   Is the star approaching or moving away?

*11-14   A star lies on the ecliptic, 90° east of the Sun on 1 March when its radial velocity is observed to be 19 km/sec toward the Earth. On 1 September its radial velocity is observed to be 40 km/sec away from the Earth. From this data

        (a)   Compute the star's radial velocity relative to the Sun.

        (b)   Compute the speed of the Earth (km/sec) in its orbit.

        (c)   Compute the value of the astronomical unit (in km).

11-15   (a)   What is the motion (speed and direction) of the Sun relative to the nearby stars?

        (b)   How has this motion been determined?

        (c)   In which direction of the sky would you expect the stars to have a general motion toward the Sun?

11-16   What three effects will cause the constellations to change their appearances over long periods of time?

11-17   (a)   Which two factors cause the position of a star (its right ascension and declination) to continuously change during the years?

        (b)   Which of the two factors is independent of the star's distance, and which is negligible for very distant stars?

11-18 (a) Cite two astronomical phenomena which can be observed with the naked eye today, and which would have appeared basically the same one million years ago.

(b) Cite two observable phenomena which would have appeared much different one million years ago.

Stellar Magnitudes and Colors

11-19 What two factors determine how bright a star appears in the sky?

11-20 Cite three ways to estimate a star's apparent magnitude.

11-21 (a) Why is a star's apparent magnitude not a good indication of the star's energy output?

(b) Why is a star's absolute visual magnitude not an accurate measure of a star's total energy output?

11-22 (a) Which kinds of stars would be expected to emit large amounts of ultra-violet radiation?

(b) Which kinds of stars would be expected to emit large amounts of infrared radiation?

11-23 Consult a table of the 20 brightest stars in the sky (found in the back of most astronomy texts).

(a) Which are also among the 20 nearest stars?
(b) How many have large proper motions (greater than 1'' per year)?
(c) Do the stars have a preference for northern or southern hemisphere?
(d) If the 20 stars were all observed at the same distance of ten persecs, which would appear faintest and which brightest?

11-24 Consult a table of the 20 nearest stars in the sky (found in the back of most astronomy texts).

(a) Are most of the stars visible or invisible to the naked eye?
(b) Which are also among the brightest stars in the sky?
(c) Do these stars tend to have large proper motions ($\mu > 1''$)?
(d) If all the stars were observed at the same distance of ten parsecs, list the stars (in their order of brightness) which would be visible to the naked eye.
(e) If all 20 stars were placed so they appeared to be equally bright, which would be nearest and which would be farthest?

11-25 The bright star Antares has apparent magnitude = 0.9, absolute magnitude = -4.5.

(a) What would be Antares' apparent magnitude as seen from 10 pc?

(b) Is Antares nearer or farther than 10 pc?

(c) Calculate the distance to Antares in parsecs and in light years.

11-26 Calculate the absolute magnitudes of the following stars:

(a) ϵ Indi   d = 3.5 pc,   m = 4.7
(b) Rigel   250 pc   0.1
(c) β Crucis   120 pc   1.4

11-27 Determine the distances (in parsecs) to each star:

(a) Ross 614   m = 11,   M = 13
(b) Deneb   1.3   -6.9
(c) Sirius   -1.4   1.4

11-28 At what distance can the brightest known stars be seen by the naked eye?

11-29 How much brighter is Rigel than the star 61 Cygni?

11-30 (a) In what way is the Sun different from all the other stars?

(b) Would the Sun be a naked eye star if it were 10 parsecs away?

*(c) What would be the Sun's apparent magnitude as seen from Pluto?

11-31 (a) Which characteristics of a star can be determined by one night's observing with the naked eye?

(b) Which characteristics of a star can be determined by one night of telescopic observations, including auxiliary instruments?

11-32 White dwarfs are believed to constitute a small but significant percentage of the stellar population, yet we observe very few of them. Why?

11-33 (a) What characteristic determines the color of a star?

(b) Put the following color stars in order of increasing surface tempera-ture: orange, white, red, blue, yellow.

11-34 Two stars of equal luminosity (energy output). Star B is blue, star Y is yellow-orange.

(a) Which has a positive color index?
(b) Which has a negative color index?
(c) Which star appears brighter to the eye?
(d) Which star appears brighter on a photo?

11-35 This question is to test your understanding of the relationships between various characteristics of a star. If the characteristic in the left column is increased, what happens to the characteristic in the right column? Does it increase, decrease, or does it remain unchanged. All other properties of the star are unchanged.

As this is increased	What happens to the numerical value of this?
distance	parallax
distance	apparent magnitude
distance	absolute magnitude
surface temperature	absolute magnitude
absolute magnitude	apparent magnitude
size	absolute magnitude

Spectral Class & the H-R Diagram

11-36   (a)   The spectral class of a star depends upon which two properties of the star?

        (b)   List the following stars in order of increasing surface temperature: A0, B3, F2, M3, G2, O8.

11-37   Determine the color and approximate surface temperature of the following stars:

Sirius	A1 spectral class
α Centauri	G2
Arcturus	K2
Rigel	B8
Betelgeuse	M2
β Crucis	B0

11-38   Three stars are observed to put out their maximum light at the following wavelengths. Estimate the spectral class of each.

        (a) 2.6×10^{-5} cm.
        (b) 5.0×10^{-5} cm.
        (c) 9.7×10^{-5} cm.

11-39   In which stellar spectral class is each of the following most likely to be found in the spectrum?

        (a)   Strong hydrogen lines,
        (b)   Lines of neutral metals,
        (c)   Lines of neutral helium.

11-40   What is the difference between three stars of spectral classes K2I, K2III, and K2V?

11-41   Estimate the absolute magnitudes of stars with the following spectral classes:
        (a)   G2V   (b)   F0II   (c)   K5III.

*11-42   (a)   Using the method of spectroscopic parallax, estimate the distance to the star 66 Pisces, of apparent magnitude = 5.7, and spectral class A1V.

        (b)   Why is this method of distance determination inaccurate?

11-43   (a)   How do astronomers estimate the chemical compositions of stars?

        (b)   How do astronomers estimate stellar rotation rates?

11-44   (a)   Which stars rotate slower, the hot O and B stars or the cool G and K stars?

        (b)   Suggest a possible reason for the large difference in stellar rotation rates.

11-45   What different kinds of information can be obtained from the analysis of stellar spectra?

11-46 (a) Which two physical characteristics of a star are usually plotted to locate the star on the H-R diagram?

(b) What two properties of the star, in fact, determine where a star is found on the H-R diagram?

11-47 Five stars of the same apparent magnitude, m = +5:

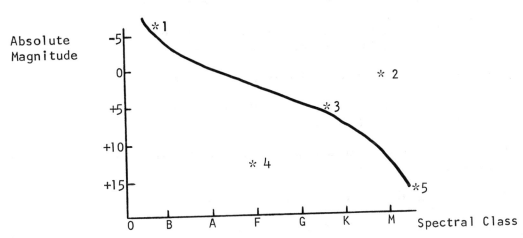

(a) Which star has the hottest surface?
(b) Which star has the coolest surface?
(c) Which star is nearest?
(d) Which star is farthest?
(e) Which are the two largest stars?
(f) Which are the two smallest stars?
(g) Which star is bluest?
(h) Which are the two red stars?
(i) Which star is most like the Sun?
(j) Which two stars don't follow the intuitive temperature-brightness relationship for stars?
(k) Which star is about 10 parsecs away?

11-48 Plot the 12 brightest stars and the 12 nearest stars on an H-R diagram, using at least half a sheet of paper. Use one symbol for the bright stars, another for the near stars. Be sure to label both axes, and indicate the location of the main sequence. Note in what general regions of the diagram the two sets of stars lie.

HINTS TO EXERCISES ON GENERAL STELLAR PROPERTIES

11-1   Distance (parsecs) = 1/parallax (arc seconds), or d = 1/p.

11-2   Measurement of parallax involves photographic astronomy.

11-3   Since d = 1/p, the smaller is p the larger is d.

11-4   Consider the relative sizes of the orbits of Earth and Mars, which would
       serve as the baselines for measuring the parallax.

11-5   d = 1/p, where p is half the major axis of the parallactic ellipse.

11-6   Assume the Sun is 1,400,000 km in diameter, and the galaxy in Andromeda is
       2,000,000 light years away.

11-7   Determination of proper motion requires photographic observations made over
       a long period of time.

11-8   The proper motion of Procyon is $1\overset{.}{.}25$ per year.

11-9   Assume the galaxy is 2,000,000 light years away. If the distance traveled
       by the star and the distance to the galaxy are in the same units, then
       distance traveled = proper motion (radians) x distance to galaxy.

11-10 Consider the answer to Exercise 11-9.

11-11 S = D x θ, where θ is in radians.
       One radian = 206265 arc seconds.
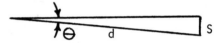

11-12 This requires observation of the star's spectrum.

11-13 H_{α} is normally observed at 6563 Å, and the speed of light is 3×10^5 km/sec.
       Also, see Exercise 3-28.

11-14

                                                        * Star

11-15 Consider the proper motions and radial velocities of stars in the solar
       neighborhood.

11-16 Consider how the positions and brightnesses of the stars change.

11-17 One factor involves the star's motion; the other is a motion of the Earth
       or of the coordinate system.

11-18 Consider the appearances and arrangements of the heavenly bodies.

11-19 Consider the star's location and energy output.

11-20 One method involves comparison with objects of known magnitudes. For another method see Exercise 4-37.

11-21 Apparent and absolute magnitudes are measures of the star's output of visible light.

11-22 Consider the way a black body radiates, as described by Wein's law (Exercise 3-26).

11-23 (d) When viewed from 10 parsecs, a star's absolute and apparent magnitude
11-24 (d) are equal.

11-25
11-26 $\log_{10} d = (m + 5 - M) / 5$, where d is in parsecs.
11-27

11-28 Assume the brightest stars are $M = -7$; the naked eye limit is $m = +6.5$.

11-29 The apparent magnitudes of Rigel and 61 Cygni are 0.14 and 5.19. Also, consider Pogson's definition of the magnitude scale.

11-30 Consider the absolute magnitude and distance of the Sun.

11-31 With the telescope you might use a photometer or a spectrograph.

11-32 Consider the ease with which stars of their brightness are seen.

11-33 Consider what happens to an iron bar as it is heated to higher and higher temperatures.

11-34 Color index = photographic (blue) mag. - visual (yellow) mag.

11-35 Review the formulas given in Exercises 11-1 and 11-25.

11-36 This is related to a star's color (Exercise 11-33).

11-37 See Exercise 11-33 for the colors.

11-38 First use Wein's law (Exercises 3-26, 10-8) to determine the surface temperatures.

11-39 The answers include spectral classes B, A, K.

11-40 The numbers I, III, and V are luminosity classes.

11-41 Star II is bright, III is a giant, and V is on the main sequence.

11-42 This is a type A1 star on the main sequence, so its absolute magnitude is about +1. Use the formula in Exercise 11-25.

11-43 Both answers involve the analysis of stellar spectra.

11-44 (b) Consider the formation of a planetary system.

11-45 Most of the physical characteristics and one of the motions.

11-46   (a)   One involves temperature or color index; the other is a measure of
             energy output.

11-47   Remember that the horizontal scale (abcissa) of the H-R diagram is also
        surface temperature.

11-48   On the H-R diagram plot absolute magnitude (visual) versus spectral class.
        For the multiple star systems (Sirius for example) plot only the brightest
        component.

ANSWERS TO EXERCISES ON GENERAL STELLAR FEATURES

11-1 $d = 1/0\rlap{.}''04 = 25$ pc $= 81.5$ LY $= 5.16 \times 10^6$ au $= 7.7 \times 10^{14}$ km.

11-2   (a)   The star is photographed every few months for several years, so that its change of position relative to the background stars is accurately determined. Then the star's parallax is numerically half the long axis of its elliptic path (shown in Exercise 11-5).

       (b)   The system of Alpha and Proxima Centauri has the largest parallax, $p = 0\rlap{.}''76$.

       (c) $d = 1/0.76 = 1.3$ parsecs.

11-3 $d = 1/0.01 = 100$ parsecs.

11-4   The size of the parallactic shifts as seen from Mars would be about 50% larger than as seen from Earth, since Mars' orbit is about 50% larger than the Earth's. Thus nearby stars could have their parallaxes measured more accurately; and some distant stars, not measurable from Earth, could be measured.

11-5   (a)   The maximum shift is in declination: $66^\circ 3\rlap{.}''5 - 66^\circ 2\rlap{.}''2 = 1\rlap{.}''3$. The parallax is then $1\rlap{.}''3/2 = 0\rlap{.}''65$. So $d = 1/0.65 = 1.54$ pc.

       (b)   The star is neither in the ecliptic plane nor at the ecliptic pole.

11-6   (a)   3.2 meters,
       (b)   129 meters,
       (c) $\sim$500 km,
       (d) 4×10^8 km, about three times the Earth-Sun distance.

11-7   (a)   The change in the position of the star relative to distant objects (stars or galaxies) is determined photographically over a long time period. Then its proper motion is the average yearly change in position.

       (b)   Barnard's star, $\mu = 10\rlap{.}''3$ per year.

       (c)   In general, the farther away the star, the smaller its proper motion.

11-8 $\frac{1}{2}^\circ/(1\rlap{.}''25/\text{yr}) = 1800''/1\rlap{.}''25 = 1440$ years.

11-9 $d = 2{,}000{,}000$ LY $= 19 \times 10^{18}$ km. $(1$ LY $= 9.5 \times 10^{12}$ km$)$
$r = 200$ km/sec $\times T(\text{sec})$
$\mu = 0\rlap{.}''01 = 4.8 \times 10^{-8}$ rad.
$r = d\mu = 200T$
$T = d\mu/200 = \dfrac{19 \times 10^{18} \times 4.8 \times 10^{-8}}{200} = 4.56 \times 10^9$ sec $= 145$ yrs.

11-10   The stars are so far away that they have shown no detectable proper motions during the brief time that man has been observing them. See Exercise 11-9, for example.

11-11 $d = 40$ pc $= 40$ pc $\times 3 \times 10^{13}$ km/pc $= 1.2 \times 10^{15}$ km.
$\mu = 0.''05$/yr $\times 4.9 \times 10^{-6}$ rad/'' $= 2.5 \times 10^{-7}$ rad/yr.
$s = d\mu = 1.2 \times 10^{15}$ km $\times 2.5 \times 10^{-7}$ rad/yr $= 3 \times 10^{8}$ km/yr $= 10$ km/sec.

11-12 (a) It is determined from the average doppler shift in the star's spectrum.

(b) About 30 km/sec.

11-13 (a) $\Delta\lambda = 6565\text{Å} - 6563\text{Å} = 2\text{Å}.$
$v = (\Delta\lambda/\lambda) \times c = (2 / 6563) \times 3 \times 10^{5}$ km/sec $= 91$ km/sec

(b) Since the wavelength is lengthened (red shifted), the star is receding.

11-14 (a) The variation in the star's radial velocity is due to the Earth's orbital motion. Thus the star's speed alone is the average $(+40 - 19)/2 = 11\frac{1}{2}$ km/sec away.

(b) $40 - 11\frac{1}{2} = 28\frac{1}{2}$ km/sec.

(c) Circumference of the Earth's orbit is
$28\frac{1}{2}$ km/sec $\times 3.16 \times 10^{7}$ sec/yr $= 9 \times 10^{8}$ km.
1 au $=$ the radius $=$ circumference/2π $= 1.43 \times 10^{8}$ km.

11-25 (a) The Sun is moving about 19 km/sec toward the star Vega in the constellation of Lyra.

(b) Determined from the study of the proper motions and radial velocities of the stars in the solar neighborhood.

(c) In the direction of Vega: $\alpha = 18$ to 19^{h}, $\delta = +35$ to $+45^{\circ}$.

11-16 Proper motions will change the relative positions of the stars. Both radial velocities and stellar evolution will change the brightnesses of the stars over long periods of time.

11-17 (a) Precession and proper motion.
(b) Precession. Proper motion is negligible for very distant stars.

11-18 (a) The physical appearances of the Sun, Moon and planets.
The phasing of the Moon, and the motions of the Moon and planets.
The rising and setting of the objects in the sky. The Milky Way.

(b) The constellations and the pole star. There were no artificial satellites.

11-19 Its distance from the Sun and its luminosity (energy output).

11-20 (1) Compare the star's magnitude with stars of known magnitude.
(2) On a photo the sizes of the stellar images can be correlated with the magnitudes of the stars. The larger the brighter.
(3) To be most accurate, use a photoelectric photometer.

11-21 (a) The star's apparent magnitude is a function of its distance and its light output.

11-21 (b) Absolute magnitude is a measure of the star's output in visible light alone, and gives no indication of its output in non-visible forms of energy such as X-rays and radio waves.

11-22 (a) Very hot O and B stars.
      (b) Very cool M stars.

11-23 (a) Sirius, α Centauri, Procyon.
      (b) Only four.
      (c) No. There are nine in the northern hemisphere (positive declination) and eleven in the southern hemisphere.
      (d) Faintest: α Centauri, Brightest: Rigel and Deneb.

11-24 (a) Only seven are visible to the naked eye.
      (b) Sirius, α Centauri, Procyon.
      (c) Yes, only two have $\mu < 1''/yr$.
      (d) At 10 pc, m = M, and Sirius, Procyon, α Centauri, τ Ceti, and ϵ Eridani would be visible.
      (e) Nearest: Wolf 359, Farthest: Sirius.

11-25 (a) At 10 pc, m = M = -4.5.
      (b) Since m is fainter than M, Antares is farther than 10 pc.
      (c) $\log d = (0.9 + 5 - (-4.5))/5 = 2.08$.
          $d = 10^{2.08}$ pc = 120 pc = 391 LY.

11-26 (a) +7
      (b) -6.9
      (c) -4

11-27 (a) 4 pc.
      (b) 540 pc.
      (c) 2.7 pc.

11-28 $\log d = (6.5 + 5 - (-7))/5 = 3.7$, $d = 10^{3.7}$ pc $\approx$ 5000 pc.

11-29 The magnitude system is defined so that two stars whose magnitudes differ by 5 will differ in brightness by 100. Thus Rigel is about 100 times brighter than 61 Cygni.

11-30 (a) Only that it is much closer to the Earth.
      (b) Yes. At 10 pc, m = M = +5, a faint naked-eye star.
      (c) $m_2 - m_1 = 5 \log (d_2/d_1) = 5 \log (39.4 \text{ au}/1\text{au}) = 8$.
          m_2 (at Pluto) = 8 + m_1 (at Earth) = 8 + (-26.5) = -18.5.

11-31 (a) Constellation, color, approximate apparent magnitude, and approximate position.

      (b) Precise apparent magnitude and position, spectral class, radial velocity, and possibly the presence of a binary companion.

11-32 They are intrinsically very faint stars, so we observe only the ones which are quite near.

11-33 (a) Its surface temperature.

      (b) Red, orange, yellow, white, blue (hottest).

11-34   (a)   Star Y,
       (b)   Star B,
       (c)   Star Y,
       (d)   Star B.

11-35   Decrease, increase, no change, decrease, increase, decrease.

11-36   (a)   Surface temperature (mainly) and surface composition.

       (b)   M3, G2, F2, A0, B3, O8 (hottest).

11-37 A1:   blue, 10,600 $^{\circ}$K;   G2:   yellow-white, 5900 $^{\circ}$K
      K2:   orange, 4700 $^{\circ}$K;   B8:   blue, 13,000 $^{\circ}$K
      M2:   red, 3200 $^{\circ}$K;   B0:   blue, 25,000 $^{\circ}$K.

11-38

	Surface temp. (Wein's law)	Spectral class
(a)	11,000 $^{\circ}$K	A0
(b)	5,800	G2
(c)	3,000	M3

11-39   (a)   A,
       (b)   K,
       (c)   B.

11-40   K2I is a bright supergiant, K2III is a giant of spectral class K2, and K2V is a lower main sequence star.

11-41   (a)   +5,
       (b)   -4.
       (c)   0, all approximate.

11-42 $\log d = (5.7 + 5 - 1)/5 = 1.94$, $d = 10^{1.94}$ pc = 87 pc.

       (b)   The absolute magnitude is only a rough average. For an A1V, M can be anywhere between 0 and +3.

11-43   (a)   Chemical composition is determined from the lines that are present in the star's spectrum. Each line indicates the presence of some element on the star's surface, although several lines may indicate the same element.

       (b)   Stellar rotation rates are also determined from the spectrum, but in this case it is the shape of the individual lines which is influenced by the rotation rate. Because of the doppler shifts in the light coming from the approaching and receding limbs of the star, rotation tends to broaden the spectral lines.

11-44   (a)   The cooler stars tend to rotate slower.

       (b)   Cooler stars have possibly lost some of their angular momentum to planetary systems.

11-45   Surface temperature and pressure, rotation speed, size, strength of magnetic fields, radial velocity, presence and relative abundances of elements, and presence of ejected shells of gas.

11-46 (a) Surface temperature, spectral class or color index versus luminosity or absolute magnitude.

(b) Age and mass.

11-47 (a) 1   (b) 5   (c) 5   (d) 1   (e) 1, 2   (f) 4, 5
(g) 1   (h) 2, 5 (i) 3   (j) 2, 4 (k) 3

11-48

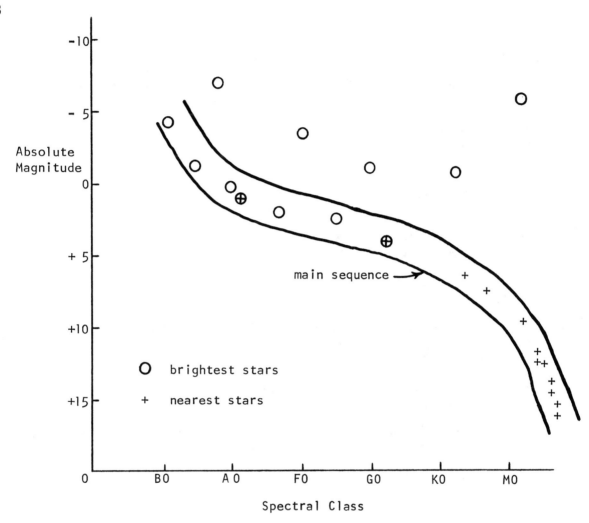

Binary Star Systems

12-1   What characteristic distinguishes each of the following kinds of binary systems?

(a) Eclipsing,
(b) Spectrum,
(c) Spectroscopic,
(d) Visual,
(e) Astrometric,
(f) Contact.

12-2   List some ways that might be used to determine if two stars which appear close together in the sky are actually an orbiting binary system.

12-3   List some ways to determine if what appears as a single star is actually a binary system.

12-4   Spectroscopic binaries appear as a single star, even as observed with the larger telescopes, and they often show the spectrum of a single star. How, then, are they identified as a binary pair?

12-5   (a) Under what conditions would we observe two orbiting stars to eclipse each other?

(b) Under what conditions would we observe no doppler shift in the spectrum of a binary system?

12-6   (a) How is it possible to have a spectroscopic binary system that is not eclipsing?

(b) How is it possible to have an eclipsing binary system that is not spectroscopic?

12-7   How does one determine if a star with a periodically changing apparent magnitude is an eclipsing binary system or an intrinsic variable star?

12-8   A visual binary system is studied by plotting the relative positions of the two stars over one period or more. Why is it that these observed relative positions do not usually represent the actual orbits of the stars?

12-9   Make a sketch of a visual binary system, and show the two position coordinates that are measured with the micrometer. Be sure to include the units of measure.

12-10 Describe three ways that the total apparent magnitude of a binary star system can vary.

12-11 Why do eclipsing binary systems typically have short periods, on the order of a few days?

12-12 To the right is the light curve of an ideal eclipsing binary system. Stars A and B are the same size, but A is brighter than B.

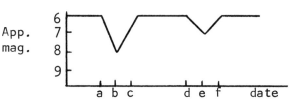

(a) At what date (a, b, ..., f) is star A totally eclipsed by star B?

(b) What is the apparent magnitude of star B?

(c) At what date is star B totally eclipsed by star A?

(d) What is the apparent magnitude of star A?

(e) Sketch the two stars (as circles) as they would appear (from Earth) at the dates a through f.

12-13 (a) Under what conditions might two stars exchange mass with each other?

(b) Under what condition would the mass exchange be only one way--the same star would always lose mass and the other always gain.

12-14 Under what condition might a binary system in which no material is exchanged evolve into a system in which the two stars do exchange material?

12-15 Explain in terms of stellar evolution how it is possible to have a binary system like Algol, which consists of a young star of 5 solar masses and an older star of only one solar mass.

12-16 What kinds of information can be learned from the study of binary star systems?

12-17 How is it known that the vast majority of the binary star systems are <u>not</u> the result of one single star having captured another single star?

Star Clusters

12-18 (a) Where in the galaxy are most of the galactic clusters found?

(b) Where in the galaxy are most of the globular clusters found?

12-19 (a) What kinds of stars are typically found in galactic clusters?

(b) What kinds of stars are typically found in globular clusters?

12-20 (a) Approximately how many galactic and globular clusters are known in the Milky Way galaxy?

(b) About how many stars are found in globular clusters and in galactic clusters?

12-21 Describe the celestial sphere as seen by an observer located

(a) In the center of a galactic cluster,

(b) In the center of a globular cluster,

(c) Near the edge of a globular cluster.

12-22 (a) Why can't we observe all the star clusters in the galaxy?

(b) Where in or around the galaxy would you expect undiscovered star clusters to be located?

12-23

Exercise 12-23. Star cluster. (Lick Observatory photo.)

The photo is of a cluster of stars which is a few hundred light years away, and whose brighter members are easily visible to the naked eye.

(a) What are the popular and descriptive names of the cluster?

(b) In what part of the galaxy are similar clusters observed?

(c) What evidence, seen in the photo, indicates that the cluster is relatively young?

(d) Why do the brighter stars appear to have points?

12-24 How do stellar associations differ from galactic clusters?

12-25 How is it determined that a given star is a member of a certain galactic cluster, assuming that both the star and the cluster are observed in the same part of the sky?

-138-

12-26 (a) How can one determine from their H-R diagrams that the stars in the
            Pleiades cluster are of a different age than the stars in the Hyades
            cluster?

       (b) Which is the older cluster?

12-27 Why do astronomers believe that globular clusters in general are much older
       than galactic clusters?

12-28 Describe the steps in the possible formation and evolution of a galactic
       cluster.

12-29 Describe in words, or with a sketch, how the locations of the center of the
       Milky Way galaxy and the Sun were determined from observations of globular
       clusters.

*12-30 The average apparent magnitude of several RR Lyrae variable stars in a
        globular cluster is m = +15. Calculate the distance to the globular cluster.

12-31 Explain how to use the H-R diagram of a cluster (galactic or globular) to
       estimate the distance to the cluster.

12-32 Why are few white dwarf stars or lower main sequence stars observed in
       globular star clusters? See Exercise 11-32.

12-33 Describe the galactic orbits of typical globular and galactic clusters.

12-34 Which kind of star cluster (galactic or globular) would you expect to be
       able to observe in or around other galaxies? Why?

12-35

Exercise 12-35.
A distant group
of stars.
(Lick Observatory
photo.)

The group of stars in the photo is about 20,000 light years from the Sun.
(a) What kind of object is it?
(b) Where in the galaxy is this kind of object found?

HINTS TO EXERCISES ON MULTIPLE STAR SYSTEMS

12-1   The characteristics involve the light curve, spectrum or position and proper motion of the system.

12-2   The methods require observations of the stars' spectra, parallaxes, proper motions or apparent magnitudes.

12-3   The methods require observations of the star's spectrum, proper motion or apparent magnitude.

12-4   Consider doppler effect.

12-5   Consider the orientation of the orbit planes of the binaries.
12-6

12-7   Consider the shapes of the light curves.

12-8   Consider the orientation of the orbit plane in space.

12-9   The two coordinates are called position angle and distance.

12-10   All three occur in close binary systems.

12-11   Consider the relationship between the period of the system, the separation of the two stars, and the chances of seeing them eclipse each other.

12-12   The Sun lies exactly in the orbit plane, so the total eclipse occurs only at the two minima of apparent magnitude.

12-13   Consider close binary systems.

12-14   Consider how a star might change its size and proximity to a companion due to stellar evolution.

12-15   There probably has been an exchange of a large percentage of the total mass of the two stars.

12-16   The information involves two basic properties of stars as well as Newton's laws.

12-17   Consider the probability of the capture process occurring.

12-18   One type of cluster is associated with the plane of the galaxy, the other with the galactic halo and nucleus.

12-19   Consider the ages of the two kinds of clusters.

12-20   Globular clusters are fewer in number, but they contain more stars.

12-21   At the center of a globular cluster it is estimated that the stars are about one astronomical unit apart, on the average.

12-22 Consider the effect of the interstellar medium.

12-23 (a) It is the Pleiades.

       (b) See Exercise 12-18 (a).

12-24 They differ in their number and kinds of stars, in addition to the distribution and motions of the stars within.

12-25 Consider the observable motions of the star and the cluster.

12-26 Look at the turn-off points on the main sequence.

12-27 Consider their H-R diagrams, and the kinds of stars they contain.

12-28 They probably begin as nebulae, and are eventually disrupted.

12-29 Consider the total volume of space occupied by all the known globular clusters in the galaxy.

12-30 Use the relationship $\log_{10} d = (m + 5 - M)/5$, where d is the distance in
12-31 parsecs, m is the apparent magnitude and M is the absolute magnitude.

12-32 Consider the large distance to a globular cluster and the brightness of those kinds of stars.

12-33 They are both in orbit around the nucleus of the galaxy.

12-34 You must be able to observe and resolve them at large, intergalactic distances.

12-35 It is a type of very old star cluster.

12-1   (a)   A single point of light (star) whose light curve shows periodic changes of apparent magnitude.

       (b)   A single point of light (star) whose spectrum is a composite of two types of spectra (A2 and G5, for example).

       (c)   The spectrum of the system shows periodic doppler shifts due to the orbital motions within the system.

       (d)   Two distinct stars whose relative positions are observed to move about each other.

       (e)   A single point of light with a wavy proper motion.

       (f)   Two stars that are so close to each other that their elongated bodies touch.

12-2   The stars' proper motions will be wavy, rather than linear, if they are a binary system. Their parallaxes will be about the same if they are a binary system. The two spectra will show different doppler shifts as the two stars orbit each other. If the stars eclipse each other, their apparent magnitudes will vary.

12-3   The spectrum may show periodic changes due to doppler shift as the two stars orbit each other. These shifts would be different from the doppler shifts due to the Earth's revolution. The apparent magnitude may periodically change if the two stars eclipse each other.

       The proper motion of the visible star may be wavy, rather than linear, if the star is in orbit about an unseen companion. The spectrum may be the composite of the spectra of two different spectral types.

12-4   The spectral lines periodically shift back and forth due to doppler shift as the two stars revolve around each other.

12-5   (a)   When the orbit plane of the two stars is oriented in space so that the Sun is very near the plane.

       (b)   When the orbit plane is oriented so as to be perpendicular to the line from the Sun to the binary system.

12-6   (a)   ⊙   ✳
                      ✳ ↖ orbit plane

       (b)   It's not possible.

12-7   The shapes of the light curves of eclipsing binaries and intrinsic variables are quite different.

12-8   Only when the orbit plane of the two stars is perpendicular to the line of sight--when the Sun is directly above or below the orbit plane--is the observed path the actual path. In most cases the orbit is inclined so that the observed path is not the real one. In the extreme case, the Sun lies in the orbit plane and the stars appear to move back and fourth in straight lines, eclipsing each other.

12-9   The position angle p is
        measured in degrees. The
        distance d is measured
        in seconds of arc.

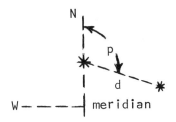

12-10   The two stars eclipse each other.

        The stars are elongated, due to mutual gravitational attractions, so that
        they present different amounts of surface area to the observer as they
        orbit each other.

        One companion may in essence reflect light from its companion just before
        conjunction, thus causing a slight brightening just before eclipse.

12-11   In order for us to observe two stars eclipsing each other, at the vast
        interstellar distances, they must be fairly close together. Otherwise
        they would be likely to pass above or below each other. According to
        Kepler's third law, if two stars are close together their period or revolu-
        tion is short.

12-12   (a) date = b,   (b) m_B = +8,   (c) date = e,   (d) m_A = +7,

        (e)

12-13   (a) If two stars are quite close together, then their tidal effects can
             gravitationally pull mass from each other.

        (b) If one close companion is a black hole, from which nothing can escape,
             then the mass would flow only from the companion into the black hole.

12-14   As a main sequence star evolves into a red giant its size increases, and
        thus its surface gets closer to any binary companion. This new proximity
        of the stars could allow an exchange of material to begin.

12-15   The star which is presently the smaller of the two was originally the more
        massive and the larger. When it first expanded to the red giant phase it
        transferred most of its mass to the star which now has the five solar masses.

12-16   That stars are probably formed as multiple systems.
        The sizes and masses of stars.
        That Newton's laws of motion and gravitation act as least within the Milky
            Way galaxy.
        That there are objects of planetary masses orbiting nearby stars.

12-17   There are far too many binary systems for them all to have been the result
        of the unlikely capture process.

12-18   (a) In or near the spiral arms, in the solar neighborhood.

        (b) They are distributed in the halo and nucleus of the galaxy, with a
             concentration toward the nucleus.

12-19 (a) Primarily young main-sequence stars, and perhaps a few giants.

       (b) Older stars: lower main sequence, red giants, Population II cepheids
            and RR Lyrae variables, sub giants, and stars on the horizontal branch
            of the H-R diagram.

12-20 (a) About 1000 galactic and 120 globular clusters.

       (b) From 100,000 to 1,000,000 stars in the globular clusters; from about
            50 to 1000 stars in galactic clusters.

12-21 (a) The sky would look similar to its appearance from Earth, except there
            would be a few hundred very bright cluster stars and perhaps some
            clouds of glowing gas in the sky.

       (b) The sky would never be dark since there would always be several "suns"
            in the sky.

       (c) Half the celestial sphere would look similar to its present appearance
            from Earth; the other half would always be bright due to the many
            nearby cluster stars.

12-22 (a) The interstellar dust prevents visual observations of distant objects
            in or near the galactic plane.

       (b) There are probably hundreds (perhaps thousands) of undiscovered galactic
            clusters in the plane of the galaxy, at distances above a few thousand
            light years from the Sun. A few globular clusters are probably in or
            near the galactic plane behind the nucleus of the galaxy.

12-23 (a) Pleiades, galactic cluster.

       (b) In the plane of the galaxy.

       (c) The patches of light around the bright stars are gas and dust, whose
            presence is indicative of young stars.

       (d) They are from light reflected off the struts which support the secondary
            mirror in the telescope.

12-24 Stellar associations generally have only a few stars, and those are hot 0,
       B and T Tauri stars. Clusters have more members, including lower main
       sequence stars and a few giants. In addition, cluster stars tend to be
       closer together and to move apart at a slower rate.

12-25 If it is possible to estimate the distances to the cluster and the star,
       then they should be about the same for cluster membership. In general,
       however, this is not possible. One must look at the proper motions and the
       radial velocities of the star and the cluster. If both properties are
       nearly identical for star and cluster, then the star in question is con-
       sidered a cluster star.

12-26 (a) The older cluster has the lower turn-off point from the main sequence.

       (b) The Hyades is the older cluster.

12-27   The main sequence turn-off points for globular clusters are generally
        lower than galactic cluster turn-offs.

        Globular clusters contain only very old stars; whereas galactic clusters
        contain mostly young objects and gas and dust.

12-28   (1)   The gas and dust in the nebula condense into many young stars, with
              much remaining gas and dust in the cluster.

        (2)   Some of the gas and dust form into planetary systems which orbit the
              cluster stars; some of it gets blown out of the cluster by the radia-
              tion pressure of the hot stars.

        (3)   The stars move apart, so the cluster appears to expand.

        (4)   The stars continue to move apart so that the cluster eventually dis-
              sipates into many individual and small multiple star systems (binaries,
              triples, etc.).

12-29   Observations of their distances and directions showed that the clusters
        occupied a spherical volume of space. It was assumed (correctly) that the
        center of this spherical volume was also the center of the galaxy, and this
        showed the Sun to be closer to the edge of the galaxy than to the center.

12-30   Since the absolute magnitude of a typical RR Lyrae variable is M = 0, then
        the distance is
$$\log d = \frac{+15 + 5 - 0}{5} = +4; \qquad d = 10^4 \text{ pc} = 10{,}000 \text{ pc.}$$

12-31   The H-R diagram of the cluster is plotted using apparent magnitude vs.
        spectral class for each star. This cluster diagram is overlayed on the
        actual H-R diagram, matching the main sequences. Then the difference in
        the two ordinates m - M is read off, and the distance is computed with the
        formula given in the hint.

12-32   At the great distances of the globular clusters, those intrinsically faint
        stars are just too faint to detect.

12-33   Globular clusters are believed to have rather elongated elliptic orbits
        with the galactic center at one focus.

        The orbits of the galactic clusters are thought to be nearly circular and
        near the plane of the galaxy. The galactic nucleus is at one focus.

12-34   Globular clusters are easily observed around many nearby galaxies because
        (a) they are intrinsically very bright, and (b) they are easy to resolve
        in the halo. Galactic clusters, on the other hand, tend to be obscured in
        the dense nebular material of the spiral arms.

12-35   (a)   Globular star cluster.

        (b)   In the halo and nucleus of the galaxy.

Observed and Physical Characteristics

13-1 Describe the basic differences between eclipsing, pulsating and eruptive variables. Cite one example of each.

13-2 List two observable differences between RR Lyrae, cepheid, and long-period (red) variables. Show the ranges of the features which differ.

13-3 How does one observationally determine if a star with changing apparent magnitude is an intrinsic variable or a totally eclipsing binary?

13-4 For each of the following stars--cepheid variable, nova, supernova, pulsar-- describe

(a) What is observed about the star's apparent magnitude during the three month period that includes the star's maximum brightness.

(b) What happened in or on the star during those three months (many years ago) to cause the observed changes in magnitude.

13-5 (a) Describe the brightness changes observed in irregular variables.

(b) Cite a possible correlation between their behavior and their locations in the galaxy.

13-6 (a) If you could observe a cepheid variable from a distance of a few astronomical units--from an orbiting planet, for example--what changes would you see in its visible appearance?

(b) What would be happening inside the star to cause the observed varia- tions in its appearance?

13-7 In addition to brightness, what other characteristics change in pulsating variable stars?

13-8 As the brightness of a pulsating variable star changes, what changes are observed in the star's spectrum?

13-9 (a) Where in the galaxy are most RR Lyrae stars found?

(b) What are the ranges of their absolute magnitude and period?

13-10 Why are pulsating variables only a small percentage of the total stellar population?

Cepheid Variables and the Distance Problem

13-11 What must be <u>observed</u> about a cepheid variable in order to be able to calculate its distance?

13-12 Cite two reasons why cepheid variables are particularly useful for determining the large distances to galaxies.

*13-13 Determine the distance to the Pop I cepheid variable having the following light curve:

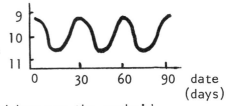

13-14 If there is an unknown interstellar cloud between the cepheid variable and the Sun--dimming the light from the star--will the calculated stellar distance be too large or too small?

*13-15 (a) A cepheid variable changes apparent magnitude by one magnitude. Its distance is 1000 pc, and it has a period of 20 days. There is no dimming by interstellar matter. Calculate the cepheid's maximum and minimum apparent magnitudes.

   (b) Calculate the cepheid's maximum and minimum apparent magnitudes if the interstellar medium dims the light by two magnitudes.

Planetary Nebulae

13-16 Explain the origin of the term "planetary nebula."

13-17

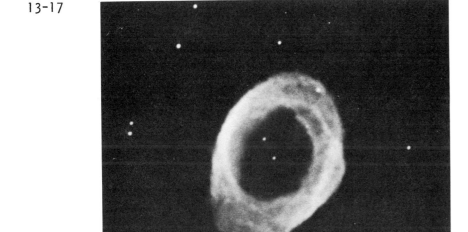

The Ring Nebula in Lyra. (Hale Observatories photo.)

The object in the photo is known as the "Ring Nebula in Lyra."

   (a) Explain what the object is.

   (b) Explain why the gaseous shell emits so much more visible light than does the hot central star.

13-18 Stars that nova and stars that become planetary nebulae both eject material into space. What evidence, however, suggests that novae and planetary nebulae are quite different phenomena?

*13-19 Calculate the distance to a planetary nebula whose shell is observed to expand $0\!''\!001$ per year in radius, and to approach the observer at 47 km/sec.

Exploding Stars

13-20 Cite some examples of the kinds of stars that eject part of their mass into the interstellar medium.

13-21 Why are the terms "nova" and "supernova" misnomers?

13-22 (a) What kinds of stars are believed to become novae?

(b) What is observed about a nova?

(c) What physically happens on a star when it becomes a nova?

13-23 (a) What kinds of stars are believed to become supernovae?

(b) What is observed about a supernova?

(c) What physically happens to a star when it becomes a supernova?

13-24 Describe the relationship between pulsars and supernovae.

13-25 (a) Describe the radiation that is observed from pulsars.

(b) What occurs on or around the pulsar to produce the observed radiations?

HINTS TO EXERCISES ON VARIABLE AND UNUSUAL STARS

13-1   Pulsating and eruptive variables are intrinsic; eclipsing variables are only apparent.

13-2   The differences are in period and spectral class.

13-3   Consider the light curves of the two systems, Ex. 12-12 and 13-13.

13-4   (a) Describe the light curves.

      (b) Except for pulsars, the changes are in the physical structures of the stars.

13-5   (b) They are observed to be in the nebular regions of the galaxy.

13-6   (a) Changes would occur in all the properties that could be observed visually.

13-7   Consider the two factors that determine a star's absolute magnitude.

13-8   The changes occur in the spectral class and in the doppler shift.

13-9   They are found in the regions thought to be occupied by old stars.

13-10 Consider the duration of the star's life during which it is a variable.

13-11 The observations are to be used to draw a light curve.

13-12 The large distances are calculated by the formula   log d = (m + 5 - M)/5.

13-13 The period and absolute magnitude of Pop I cepheids are related by:

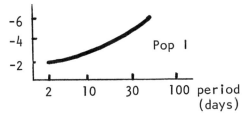

13-14 The cloud makes the variable appear fainter than it would be just due to distance. The distance is calculated by the formula cited in the hint to Exercise 13-12.

13-15 The formula is cited in the hint to Exercise 13-12. The quantity m would be the average apparent magnitude.

13-16 Consider the object's visual appearance seen through a small telescope.

13-17 Consider the kinds of radiation emitted by the hot central star, and what effect that radiation has on the shell of gas.

13-18 Consider the speeds of the ejected material and the durations of the two events.

13-19 The formula to use is $p = 4.74 \mu/V_r$, where p is the parallax and μ is the yearly rate of expansion (both in seconds of arc). The speed of approach is V_r, in km/sec.

13-20 Most stars do this, some leisurely and some violently.

13-21 Consider the original meanings of the terms.

13-22 These events occur near the end stages of some stars.
13-23

13-24 One is believed to be the remaining core of the other.

13-25 The term pulsar was given originally as a description of the way the radiation is received.

SOLUTIONS TO EXERCISES ON VARIABLE AND UNUSUAL STARS

13-1   Eclipsing stars are two that periodically eclipse each other as seen from Earth. Algol in Perseus is a well known example.

Pulsating stars periodically change their size, expanding and contracting. The δ star in Cepheus and the RR star in Lyra are examples.

Eruptive variables exhibit sudden (usually unpredicted) outbursts of light. Nova Herculis and Nova Aquillae are examples.

13-2

	RR Lyrae	Cepheid	Long Period (red)
Period (days)	0.1 to 1	2 to 50	90 to 700
Spectral class	AO to F5	F and G	M, also S,R,N

13-3   The light curves of the two kinds of systems are quite different in shape. Also, the doppler shift records are different.

13-4   Cepheid's apparent magnitude varies between maximum and minimum every few days, because the star is pulsating.

A nova suddenly gets very bright in a day or two, then gradually dims over a month or two, because the star has blown off an outer layer of gas.

Supernova suddenly gets quite bright in a day or less, then dims over two or more months, because the star has exploded.

A pulsar appears to emit several pulses of visible and radio radiation each second, because the rapidly rotating star radiates in a beam which periodically sweeps by the Earth.

13-5   (a)   Irregular variable stars, such as T Tauri and R Coronae Borealis, tend to vary their brightness slowly and erratically over months or years of time.

(b)   Since these stars are located in regions that are rich in interstellar gas and dust, and since they lie just above the main sequence on the H-R diagram, it is believed that they are very young, condensing stars, varying erratically as they interact with the interstellar medium.

13-6   (a)   Periodic changes would be observed in the star's size, apparent magnitude and possibly its color.

(b)   The observed changes are believed to be due to internal instabilities and unbalanced forces within the star. It is trying to adjust to new internal energy sources.

13-7   Changes are found in their surface temperature, spectral class, size, magnetic field, and occasionally period of variability.

13-8   The spectral class goes from an early type (say F) at maximum brightness to a later type (say G) at minimum brightness. Also, the spectrum is doppler shifted toward the red as the star contracts and toward the blue as it expands.

13-9   (a)   Most are found in the nucleus and halo of the galaxy and in the globular clusters.

         (b)   The periods range from 0.1 to about 1 day, the absolute magnitudes from 0 to +1.

13-10   The pulsations are the result of instabilities which occur during very short intervals in the life of a star. Thus at any given time only a few stars are at that stage of evolution.

13-11   You must observe its light curve: apparent magnitude vs date. Also, you must determine from its location whether it is Population I or II.

13-12   Their absolute magnitudes can be estimated from their periods. They are intrinsically very bright stars which can be seen at large distances.

13-13   The average m = +10.
The period = 30 days; therefore, M = -5.

$$\log d = (m + 5 - M)/5 = (10 + 5 - (-5))/5 = 4$$
$$d = 10^4 \text{ parsecs.}$$

13-14   The star's calculated distance is too large.

13-15   (a)   If the period is 20 days, then the absolute magnitude M = -4 from the graph in the hint to Exercise 13-13.

$$\log d = (m + 5 - M)/5, \text{ where } m \text{ is the average apparent mag.}$$
$$\log 1000 = (m + 5 - (-4))/5; \; m = +6.$$
$$m_{min} = +6\tfrac{1}{2}; \; m_{max} = +5\tfrac{1}{2}.$$

         (b) $m_{min} = +8\tfrac{1}{2}; \; m_{max} = +7\tfrac{1}{2}.$

13-16   Historically any vague, diffuse unknown object was called a nebula. The term "planetary" was used because they look similar to Neptune and Uranus as seen with a small telescope.

13-17   (a)   It is a planetary nebula, a very hot star surrounded by a shell of gas which was ejected by the hot star.

         (b)   The hot star emits large quantities of ultraviolet and X radiation which is not visible. This radiation, in turn, excites the gaseous shell which then re-emits the radiation as visible light.

13-18   The central stars are different. Also, the nova ejects its material at higher speeds than the planetary nebula, so the shell around the nova dissipates at a much faster rate.

13-19   The parallax p = 4.74 x 0".001 / 47 km/sec $\approx$ 0".0001.

The distance d = 1/p = 1/0.0001 = 10,000 pc.

13-20 Supernova, nova, flare star, planetary nebula, the Sun.

13-21 Early people believed they were seeing new stars ("novae") just turning on; whereas we now know that the nova and supernova stages occur near the end of the stars' lives.

13-22 (a) They are old stars of low mass (less than $3m_\odot$) and of spectral class A or B.

       (b) It suddenly becomes very bright, and then it slowly fades back to its previous brightness.

       (c) The star ejects its outer layer into the interstellar medium; then it returns to being an ordinary star with less mass.

13-23 (a) Stars of mass from about $3M_\odot$ to about $8m_\odot$, which have left the main sequence.

       (b) They suddenly become very bright, then slowly fade leaving glowing remnants behind.

       (c) The star explodes, ejecting much of its mass into space as glowing nebular material. The core of the exploded star is imploded to a rapidly rotating neutron star, a pulsar.

13-24 Pulsars are believed to be the imploded cores of exploded stars (super-novae), which remain as rapidly rotating neutron stars.

13-25 (a) Both radio and visible radiations are received in very rapid, short, regular pulses.

       (b) It is believed that the neutron star radiates energy in a beam, so each time the beam from the rapidly rotating star sweeps by the Earth we receive a signal.

Star Formation

14-1

Enlarged section
of the Rosette
gaseous nebula in
Monoceros, NGC
2237. (Hale
Observatory photo.)

The photo is of an enlarged portion of the Rosette gaseous emission nubula, NGC 2237.

(a) Why do astronomers believe that new stars are presently being formed in nebular regions like this?

(b) Locate two or three features in the photo which may be forming new stars.

14-2 What evidence suggests that a nebula may eventually condense into a cluster of stars, which in turn may eventually dissipate into individual or binary stars?

14-3 (a) What is a proto-star?

(b) Where in the galaxy are likely places to search for proto-stars?

(c) What kind of radiation would a proto-star be expected to emit?

14-4   Why do astronomers believe that T Tauri stars are in the pre-main sequence stage of evolution?

14-5   (a)   How do main sequence stars contribute to the content of heavy elements in the interstellar medium?

       (b)   How do other kinds of stars contribute heavy elements to the interstellar medium?

14-6   (a)   What are the basic observational differences between Population I and Population II stars?

       (b)   How are these differences explained in terms of star formation and stellar evolution?

14-7   Why is it that we can know so much about the formation of stars, which are so far away, whereas we know so little about the formation of planets, one of which is just outside the door?

Stellar Energy Production

14-8   What is the predominant energy source for each of the following kinds of stars?

       (a)   Pre-main sequence.   (c)   Red giant.
       (b)   Main sequence.   (d)   White dwarf.

14-9   (a)   State the overall effect of the proton-proton reaction.

       (b)   Where in main sequence stars does this reaction occur, and why there?

       (c)   What determines whether the proton-proton reaction or the carbon cycle is the major energy source in a given star?

14-10 What is the form or forms of the energy produced by the proton-proton reaction?

14-11 Once the proton-proton reaction begins in the core of a star, why isn't all the hydrogen converted to helium instantly?

14-12 (a)   State the overall effect of the carbon cycle.

       (b)   How does the carbon cycle differ from the proton-proton reaction?

       (c)   Where in main sequence stars does this reaction occur?

14-13 When hydrogen is converted to helium plus energy inside stars, where does the energy come from?

*14-14 A main sequence star of 20 solar masses is about 10,000 times more luminous than the Sun. Assuming the star to be pure hydrogen, how long would it take for the star to convert all its hydrogen to helium?

14-15 Cite two hypotheses that have been given historically for the way the Sun produced its energy, and mention what is wrong with each hypothesis.

14-16 When a main sequence star exhausts its hydrogen fuel in its core,

    (a) What happens to the core of the star?

    (b) What happens to the star's outer layers?

14-17 (a) What two reactions are believed to occur in red giant stars, and where in the star do they occur?

    (b) Why don't both these reactions occur in main sequence stars?

14-18 How do we know that stars cannot produce energy forever?

14-19 (a) What is the meaning of Einstein's famous formula $E = mc^2$?

    (b) How does the formula apply to stellar energy production?

Evolution and the H-R Diagram

14-20 Arrange the following in order of increasing age (youngest first)

    (a) Star in the Orion nebula,   (d) The universe,
    (b) Star in the Pleiades,   (e) A meteor,
    (c) The oldest living thing on Earth,   (f) The Sun.

14-21 On an H-R diagram, show where each of the following kinds of stars are located:

    (a) Converts H to He in its core.
    (b) Converts H to He in an envelope surrounding the core.
    (c) Very young massive stars.
    (d) Stars that pulsate with regular periods.
    (e) Converts He to C in its core.
    (f) Very old and very dense.
    (g) Ejecting its outer layers into space.
    (h) Proto stars.

14-22 What does a star do in its evolution just <u>after</u> each of the following events or stages? What is the next evolutionary step?

    (a) Reaches the zero-age main sequence.

    (b) Ejects material as a nova.

    (c) Converts most of the hydrogen in its core to helium.

14-23 (a) What property of a star determines where it is located on the main sequence?

    (b) What is the basic difference between upper and lower main sequence stars?

14-24 (a) When a star changes its position on the H-R diagram, what does this indicate is happening inside the star?

    (b) Why do we not regularly observe stars changing their positions on the main sequence?

14-24 (c) Why can't we observe the main sequence against a clear dark sky?

14-25 Outline on an H-R diagram the evolutionary track of a star twice as massive as the Sun, beginning with the proto-star. Be sure to label the axes and indicate all specific stages such as red giant, main sequence, white dwarf, etc.

14-26 The stars on the main sequence of the H-R diagram follow the intuitive expectation that hot stars are bright and cool stars are faint. How do we explain, then, the bright, cool red stars on the upper right portion of the diagram and the faint, hot white stars on the lower left side?

14-27 Describe the basic difference between:

   (a) A main sequence A0 star and a white dwarf A0 star.

   (b) A main sequence M2 star and a red giant M2 star.

14-28 (a) About how much time has the Sun spent on the main sequence, and about how much longer will it stay on the main sequence?

   (b) What major changes have occurred in the Sun, if any, since it arrived on the main sequence?

14-29 (a) Why is it believed that stars spend the major portion of their lives on the main sequence?

   (b) Why do upper main sequence stars (0 and B) spend much less time on the main sequence than do lower main sequence stars?

14-30 (a) How can we tell from their H-R diagrams that the stars in the Pleiades cluster are of different age than the stars in the Hyades cluster?

   (b) Which of the two clusters is older?

14-31 (a) Sketch the H-R diagrams for a typical globular cluster and a typical galactic (open) cluster.

   (b) Explain why the diagrams differ. In particular, in globular clusters why do we observe very few upper and lower main sequence stars or very few white dwarfs?

Stellar Fates

14-32 (a) What are three possible end stages for stars?

   (b) What property of a star determines how it will end?

   (c) What will be the fate of the Sun?

14-33 (a) How is it known that white dwarfs are dying stars, and not stars in some active stage of evolution?

   (b) Why is it difficult to observe white dwarfs?

14-34 What are neutron stars, and how are they detected?

14-35 (a) Why is it difficult to detect black holes in space, if they exist?

(b) If black holes do exist but emit no electromagnetic radiation, then how might they be detected?

(c) What is one likely candidate for a black hole?

14-36 Describe the fate of a planet orbiting a star as that star collapses to become a black hole. Specifically, what will happen to the planet's orbital motion and energy sources?

14-37 Why doesn't a proto-star just continue to collapse until it becomes a planet or a black hole?

14-38 How do we know that planets are not dead stars?

HINTS TO EXERCISES ON STELLAR EVOLUTION

14-1   (b) New stars are thought to appear as small dark "globules" during part of their early formation.

14-2   Consider the appearance of a young cluster like the Pleiades and the motions of the stars within a young cluster.

14-3   See Exercise 14-1, and review black body radiation from cool bodies.

14-4   One reason is their positions on the H-R diagram; the other is their locations in space. Also, see Exercise 13-5.

14-5   See Exercise 10-32.

14-6   Pop I stars are believed to be relatively young, and Pop II stars are relatively old.

14-7   There are two reasons: one involves the present state of the planetary surfaces; the other the numbers of the objects in question.

14-8   Each uses a different process, but one does not produce energy.

14-9   This reaction involves only hydrogen (H) and helium (He).

14-10   It is in several forms: radiation, particles and motion.

14-11   At a given instant in any gas, a few of the molecules are moving at very high speeds, a few are moving at very low speeds, and most are moving at intermediate speeds.

14-12   This reaction involves hydrogen, helium, carbon, oxygen and nitrogen.

14-13   See Exercise 14-19 which describes the relation $E = mc^2$.

14-14   The Sun's mass and luminosity are 2×10^{33} grams and 4×10^{33} ergs/sec. When one gram of H is converted to He, 6.4×10^{18} ergs of energy are produced.

14-15   It is believed that the Sun has been shining much as it is today for over four billion years.

14-16   It is believed that main sequence stars evolve into red giants.

14-17   The reactions involve hydrogen, helium and carbon. One reaction occurs in the envelope surrounding the star's core.

14-18   See Exercise 14-19.

14-19   It is believed that stars convert mass to energy in their cores.

14-20   Be sure you understand what a meteor is . See Exercise 9-24.

14-22 See Exercise 14-16.

14-23 The two basic properties of a star are its age and mass.

14-24 A star produces energy by several different processes during its lifetime, and some processes may continue for billions of years.

14-25 This star will be higher on the main sequence than the Sun, and it will evolve faster.

14-26 A star's intrinsic brightness is determined by its surface temperature and its size.

14-27 The differences involve the stars' sizes and energy sources.

14-28 See Exercises 10-40 and 10-41.

14-29 Generally, the more massive a star is the faster it consumes its fuel.

14-30 It is believed that all the stars in a given cluster were formed at about the same time, and that the more massive stars evolve faster than the less massive stars in the cluster.

14-31 Globular clusters are believed to consist of very old stars; whereas galactic clusters contain relatively young objects.

14-32 The choice of a star's ultimate fate is based upon its initial mass.

14-33 Consider their size, mass and luminosity.

14-34 They are one of the three possible end stages of stars.

14-35
14-36 Black holes are theorized to be collapsed, massive stars with such high surface gravity that no electromagnetic radiation escapes the object.

14-37 Consider the effect of energy production starting in the core of a collapsing proto-star.

14-38 Consider the differences in the properties of planets and of dead or dying stars, specifically the masses and densities.

SOLUTIONS TO EXERCISES ON STELLAR EVOLUTION

14-1   (a) New stars are believed to form out of gas and dust, and the nebulae are made of gas and dust. Also, very young stars are observed in these regions and a few have been seen to turn on.

       (b) The small dark spots within the bright regions are possible forming stars.

14-2   Very young stars are believed to condense out of nebular material, and the large amount of gas and dust surrounding the stars in the Pleiades cluster suggests that it was once a nebula. Within galactic clusters the stars move away from each other--the cluster expands--which suggests that these clusters eventually dissipate.

14-3   (a) A cloud of gas and dust which is collapsing to become a star, and which is just beginning to radiate.

       (b) The nebular regions of the galactic plane.

       (c) Since it is relatively cool, it would initially radiate infrared and microwaves.

14-4   On the H-R diagram they lie just above the main sequence, where theory predicts pre-main sequence stars to be. Also, they are observed in nebular regions believed to be the birth places of stars.

14-5   (a) Particles are ejected from their surfaces, like the solar wind and solar flares.

       (b) Novae and supernovae eject substantial percentages of their mass into space.

14-6   (a) The spectra of Pop I stars have more lines due to heavy elements; whereas Pop II spectra show mostly H and He lines. Also, Pop I stars are associated with the spiral arms of the galaxy, while Pop II stars are halo and nucleus objects.

       (b) It is thought that the old Pop II stars formed when the universe was young, when the universe consisted of mostly H and He. Then these stars produced heavy elements through nuclear fusion, exploded (supernova), and distributed the heavy elements into the interstellar medium. The younger Pop I stars were then formed out of the material that included remnants of the older stars, the heavy elements.

14-7   First, erosion has so changed the surfaces of the planets that no evidence of their formation remains. Second, there are billions of stars to examine, in all stages of evolution; but there are only nine samples of planets and and a few satellites, all of which are probably the same age.

14-8   (a) Conversion of gravitational potential energy into heat.

14-8   (b)   Conversion of hydrogen into helium plus energy.

       (c)   Conversion of helium into carbon plus energy.

       (d)   Shines by cooling off, no longer producing energy.

14-9   (a)   Four H nuclei combine to produce one He nucleus plus energy.

       (b)   In the cores of the stars, because it is only there that the pressure and temperature are high enough for the reaction to take place.

       (c)   The central temperature of the star. If it is above about 15 million degrees K, then the carbon cycle produces most of the energy; below that the proton-proton reaction predominates.

14-10 Neutrinos and gamma rays, both forms of energy, are emitted. Positive electrons (β particles) are emitted and quickly combine with negative electrons to produce gamma rays. Also, the final product (He) is at a higher energy than the initial protons.

14-11 At a given instant only a few protons have sufficient energies to overcome the repulsive forces of their positive charges and get close enough to react. Most of the protons have too little energy to react.

14-12 (a)   Four H nuclei combine to produce one He nucleus plus energy. Carbon serves as a catalyst to the reaction.

       (b)   The proton-proton reaction occurs at lower temperatures and requires no carbon catalyst.

       (c)   In the hot cores of the stars.

14-13 The resulting helium nucleus has slightly less mass than the four protons which react to form it, and the mass which is "lost" is converted into energy according to $E = mc^2$.

14-14 When the entire star is converted into He it will produce

       $E = 40 \times 10^{33}$ grams $\times$ 6.4×10^{18} ergs/gram $= 256 \times 10^{51}$ ergs.

       Since the star's output $L = 10,000\ L_\odot = 4 \times 10^{37}$ ergs/sec, the time required will be

       $$T = \frac{E}{L} = \frac{256 \times 10^{51}\ \text{ergs}}{4 \times 10^{37}\ \text{ergs/sec}} = 64 \times 10^{14}\ \text{sec} \approx 200\ \text{million years.}$$

14-15 Red hot stone (Socrates), burning coal, gravitational contraction, and the combination of hydrogen and oxygen to produce water plus energy have all been proposed. The problem is that none of these can sustain the Sun's present energy output for even one billion years, let alone four billion years.

14-16 (a)   The core collapses and heats up until the helium combines to produce carbon plus energy.

       (b)   The outer layers expand and cool.

14-17 (a) Helium nuclei are converted into carbon nuclei ("triple alpha" process) in the cores. The carbon cycle converts H into He plus energy in the envelope that surrounds the core.

(b) The cores are not hot enough.

14-18 A star derives its energy by the conversion of its mass into energy according to $E = mc^2$. Even if a star could convert all its mass into energy (and no star can), eventually it would use up all its mass and thus all its energy.

14-19 (a) When a substance of mass m is converted into energy E, then the amount of energy which is produced is $E = mc^2$, where c is the speed of light.

(b) It is thought that stars convert very small amounts of their mass into tremendous amounts of energy by this process.

14-20 e (youngest), c, a, b, f, d (oldest).

14-21 (a) All along the main sequence.

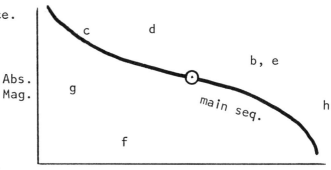

14-22 (a) Spends most of its life on the main sequence.

(b) Collapses to become a white dwarf.

(c) The core collapses and heats up, so that He is converted to carbon plus energy. The star becomes a red giant.

14-23 (a, b) The larger the initial mass of the star, the higher it is on the main sequence.

14-24 (a) The star is changing from one energy source to another.

(b) The changes in the energy processes which cause the star to change position on the H-R diagram take much longer than the hundred years or so during which man has closely studied the stars.

(c) The main sequence is not a place in space; it is merely a region on a diagram.

14-25

Abs.
Mag.

variable

red giant

nova

main seq.

white
dwarf

proto-star

Spectral Class

14-26 The bright, cool red stars are the very large red giants, which derive
their brightness from their large size. The faint, hot white stars are
the small white dwarfs.

14-27 (a) The main sequence A0 star is larger, more massive, and is producing
energy in its core. The white dwarf is very small and is producing
no energy.

(b) The red giant is much larger and more massive. Most of its energy
is produced by the triple alpha process; whereas the small main
sequence star gets energy from the proton reaction.

14-28 (a) Both answers are about 5 billion years.

(b) Very little change has occurred.

14-29 (a) In a random sample of stars, most of the stars are found to lie on
the main sequence.

(b) The very massive upper main sequence stars emit tremendous amounts
of energy, and to do this they must use their available H very
quickly and then evolve to the red giant stage. The small lower
main sequence stars, on the other hand, emit much less energy and
thus use their H very slowly.

14-30 (a) Their H-R diagrams are much different, because in the older Hyades
cluster most of the upper main sequence stars have had enough time
to evolve to the red giant stage.

(b) In the younger Pleiades cluster the turn-off point is higher, indicat-
ing that its stars have not had as much time to evolve off the main
sequence.

14-31 (a)

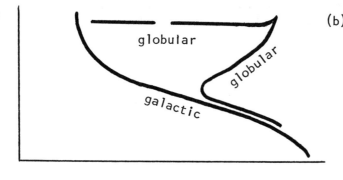

globular

globular

galactic

(b) In the older globular
cluster most of the
stars have had time to
evolve to the red giant
stage and beyond. In the
young galactic cluster
only the upper main
sequence stars are moving
toward the red giant
stage, and the lower main
sequence stars are still

14-31 (b) (Con't)
      evolving to the main sequence. Few white dwarf stars are observed
      in globular clusters because they are too faint to detect at the great
      distances of the globular clusters.

14-32 (a, b) The star's initial mass determines its fate:

      $m < 3.2m_\odot$ white dwarf

      $3.2m_\odot \leq m \leq 8m_\odot$ supernova and pulsar, neutron star

      $8m_\odot < m$ black hole

      (c) A white dwarf.

14-33 (a) Theory predicts that a star of one solar mass (typical of a white
          dwarf) with a nuclear energy source would be much larger than a
          typical white dwarf (10,000 km diameter) and would emit more energy
          than do white dwarfs.

      (b) They are intrinsically faint stars and therefore difficult to observe
          at great distances.

14-34 They are believed to be the remnant cores of stars that have exploded
      (supernova) and are composed almost entirely of densely packed neutrons.
      They are detected either by their very rapid, regular radio pulses, or
      by being at the center of a supernova.

14-35 (a) They neither emit nor reflect any electromagnetic radiation, and so
          cannot be observed in the usual manner of other stars.

      (b) Look for a non-visible binary star which emits X rays and which is
          at least five times as massive as the Sun.

      (c) The X ray source Cygnus X-1 is believed to have all of the character-
          istics of a black hole.

14-36 The planet will continue to move in its same orbit, since the planet (at
      some distance from the collapsing star) continues to feel the same gravita-
      tional attraction as it did before the collapse. On the other hand, the
      planet would no longer receive any radiation from the star, so that all its
      energy would have to be derived from sources on or in the planet.

14-37 When the core of the collapsing proto-star heats up enough, nuclear reac-
      tions produce energy. The force of the energy moving from the hot core
      to the cooler surface just balances the gravitational forces that would
      collapse the star, so the collapse is halted.

14-38 Stars (even dead ones) are more massive than even the most massive planets,
      and the matter in dead or dying stars is thousands of times more dense than
      planetary matter.

Shape, Structure and Motions

15-1   (a)   Describe, with a sketch, the size and shape of the Milky Way galaxy.

     (b)   About how many stars does it contain?

     (c)   Is the Milky Way a common or unusual kind of galaxy?

15-2   (a)   Why was it long suspected that the galaxy had a spiral structure?

     (b)   How was this spiral structure first confirmed?

     (c)   Describe the role of radio astronomy in mapping the structure of the galaxy.

15-3   How was the true location of the Sun in the galaxy first determined?

15-4   What could you conclude about the shape of the galaxy and the Sun's location in it if you observed that

     (a)   The stars were uniformly distributed all around the celestial sphere, as observed by any sized telescope?

     (b)   The stars were uniformly distributed all around half the celestial sphere, and the other half showed distant galaxies but no individual stars?

     (c)   The milky way were a band of light extending only half way around the celestial sphere.

15-5   (a)   As seen with the naked eye, what is the observed distribution of stars on the celestial sphere?

     (b)   As seen through a small telescope, what is the distribution of stars on the celestial sphere?

     (c)   Why are the answers to (a) and (b) different?

15-6   (a)   What constellation is observed in the direction of the center of the galaxy--in the direction of $0°$ galactic longitude?

     (b)   Why isn't that the brightest region on the celestial sphere, as seen with the unaided eye?

     (c)   What naked-eye feature identifies the approximate location of galactic latitude $0°$ on the celestial sphere?

15-7 (a) What kinds of evidence indicate that the galaxy is rotating?

(b) What is the motion of the Sun within the galaxy?

15-8 How does the rotation of the galaxy affect the average proper motions and radial velocities of the stars in the neighborhood of the Sun? Use a sketch to illustrate.

15-9

The galaxy NGC 4565. (Hale Observatories photo.)

The photograph is of the galaxy NGC 4565, which is believed to appear as the Milky Way would if observed edge-on. Assume that the galaxy is the Milky Way and point out the locations of the nucleus, the halo, the spiral arms, and the Sun.

15-10 Describe the kinds of objects that are found in the Milky Way galaxy's

(a) Spiral arms,
(b) Halo,
(c) Nucleus.

15-11 Cite two ways that we study the nucleus of the galaxy.

*15-12 Estimate the mass of the Milky Way galaxy, using Newton's modification of Kepler's third law and the Sun's revolution period around the galaxy.

15-13 (a) Which are the naked-eye objects on the celestial sphere that are <u>not</u> members of the Milky Way galaxy?

(b) Which of the above can be observed from the United States?

Nebulae and the Interstellar Medium

15-14 (a) Why can't we observe light from the center of the galaxy?

(b) What kind of radiation can be detected from the center of the galaxy, and why?

15-15 (a) What are some observable phenomena which are caused by the inter-
            stellar medium?

      (b) In what part of the galaxy is most of the dust located?

15-16 (a) What is an emission nebula?

      (b) What color is often associated with one, and why?

      (c) Where are emission nebulae usually found in the galaxy?

      (d) Name one emission nebula.

15-17 Even though hydrogen is the most abundant element in the universe, other
      elements often show stronger emission lines in some nebulae. Explain this.

15-18 (a) What is a reflection nebula?

      (b) What is the characteristic color of a reflection nebula, and why?

      (c) Where are they usually found in the galaxy?

      (d) Name a well-known reflection nebula.

15-19 (a) What is a dark nebula?
      (b) Where in the galaxy are most dark nebulae found?
      (c) Name a well-known dark nebula.

15-20 Describe what must happen in or near a dark nebula in order for it to
      evolve into a bright emission nebula over, say, the next million years.

15-21 (a) How is the presence of interstellar gas detected in the galaxy?

      (b) Where is most of the interstellar gas located in the galaxy?

15-22 (a) What is the relationship between an HII region and an emission nebula?

      (b) Interstellar dust is often found in emission nebulae; why is it rarely
           found in HII regions?

15-23 (a) Explain how and why the composition of the interstellar medium has
           slowly changed during the life of the galaxy.

      (b) What evidence do we have for the answer of part (a)?

15-24 Starlight that comes to us from large distances is sometimes polarized.
      What causes this polarization?

15-25 What evidence do we have for the existence of large-scale magnetic fields
      within the Milky Way galaxy?

15-26 (a) What is 21-centimeter radiation, and how is it produced?

      (b) How is it observed or detected?

      (c) How was its presence in the galaxy first discovered or predicted?

      (d) Observations of 21-cm radiation indicate what?

15-27 What are "forbidden lines" in a spectrum, and why do they in fact occur?

15-28 (a) List some of the organic molecules which have been detected in space.

       (b) How have they been discovered?

15-29

Exercise 15-29. The Horsehead Nebula in Orion, NGC 2024. (Hale Observatory photo.)

The photo, of the Horsehead Nebula in the constellation of Orion, shows three kinds of objects:
        1. Dark nebular material,
        2. Stars,
        3. Glowing gas (emission nebula).

(a) On the photo, identify the three kinds of objects.

(b) Which of the three is closest to the Sun; which is farthest?

Relationships of Stars and Galactic Features

15-30 If you wished to observe stars being formed or created, in what part of the galaxy would you look, and why there?

15-31 Describe the two main stellar populations in terms of

       (a) Their basic differences.

       (b) The kinds of objects in each population.

       (c) Where in the galaxy the two populations are found.

15-32 Suggest a reason why younger stars have greater abundances of the heavier elements than do the relatively older stars.

15-33 (a) What are the so called "high velocity" stars?

(b) How do their orbits differ from the solar orbit?

15-34 It is estimated that about 25 novae per year occur within the Milky Way galaxy.

(a) Why do most of these novae go undetected?

(b) Why do we expect that this rate ($\sim$25/yr) was lower in the distant past when the galaxy was younger?

15-35 (a) Cite some sources of infrared radiation in the galaxy.

(b) Cite some sources of ultraviolet radiation in the galaxy.

(c) How are infrared and ultraviolet sources detected?

15-36 (a) Cite some objects within the galaxy which are sources of X-ray radiation.

(b) How are these sources distributed within the galaxy?

(c) How are X-ray sources observed?

15-37 (a) Cite some sources of gamma radiation within the galaxy.

(b) How are they observed?

15-38 (a) What are cosmic rays?

(b) Where is it believed that they originate?

Extra-Terrestrial Intelligent Life in the Galaxy

15-39 List some reasons why scientists suspect that life is not unique to the planet Earth.

15-40 Why is it unlikely that the Earth is presently being visited by extra-terrestrial beings?

15-41 (a) If we wished to "advertise" our presence within the community of suspected galactic civilizations, how might we go about doing this?

(b) How is it that nearby advanced civilizations (if they exist) may already know of our existence?

15-42 What kinds of stars are thought to be the best candidates for having life-supporting planets in orbit around them, and why?

15-43 What would be some of the possible advantages and disadvantages for the United States should we attempt to search for and communicate with extra-terrestrial civilizations that may live around some of the nearby stars?

HINTS TO EXERCISES ON THE MILKY WAY GALAXY

15-1   (c) It is similar to the great galaxy (M 31) in Andromeda.

15-2   Consider observations of (a) other galaxies thought to be similar, (b) bright emission nebulae which can be seen at large distances, and (c) 21-cm radiation from neutral hydrogen.

15-3   Observations of the distribution of globular clusters were used.

15-4   Consider both spherical and disk-shaped galaxies.

15-5   Consider the appearances of the stars in the milky way as observed by the two methods.

15-6   (b) Consider the effect of the interstellar dust.

15-7   The laws of orbital motion apply to stars orbiting the galactic nucleus as well as to the planets orbiting the Sun.

15-8   The stars are assumed to orbit the nucleus and move according to Kepler's laws of motion.

15-9   The Sun is located in the spiral arms, about two-thirds of the way out from the nucleus.

15-10   The younger objects and the material from which they are formed populate the spiral arms; whereas the halo and nucleus contain older objects.

15-11   We do not receive visible light from the Milky Way's nucleus.

15-12   Kepler's third law was modified to read

$$(M_{galaxy} + M_{Sun}) \, P_{Sun}^2 = a_{Sun}^3,$$

where the two masses are in terms of the Sun's mass ($M_{Sun} = 1$), the Sun's period is about 200,000,000 years, and a, the distance from the Sun to the galactic center, is 2×10^9 au.

15-13   They are all nearby galaxies.

15-14   See Exercise 15-6b.

15-15   Consider the nebulas.

15-16   (a) It is a region of hot, glowing gas.

15-17   Many electron transitions radiate invisible ultraviolet or infrared radiation.

15-18   These are usually seen as bluish areas near hot stars.

15-19 These are regions which appear unusually dark.

15-20 Review what is an emission nebula and what causes it to shine.

15-21 (a) It radiates at various wavelengths, depending on its temperature.

15-22 HII is ionized hydrogen--hydrogen atoms that have lost their electrons.

15-23 Consider the effect of the stars which return part of their gases to the interstellar medium.

15-24 Consider galactic magnetic fields and interstellar dust.

15-25 See Exercise 15-24.

15-26 It is electromagnetic radiation in the microwave or radio part of the spectrum.

15-27 They are spectral lines which cannot be created on the Earth, neither naturally nor in the laboratory.

15-28 They emit very faint radio signals.

15-29 The dark nebular material is opaque to visible light.

15-30 It is believed that stars initially condense from gas and dust.

15-31 One consists of objects that are thought to be relatively young, the other of older objects.

15-32 Consider the compositions of the interstellar materials from which the two kinds of stars were formed. See Exercise 15-23.

15-33 Compare the orbits of the high-velocity stars to the Sun's orbit.

15-34 (a) Consider the effect of the interstellar medium.

(b) Consider the evolution stage of a nova.

15-35 Only small amounts of ultraviolet and infrared radiation penetrate the Earth's atmosphere.

15-36 (c) X-ray radiation does not penetrate the Earth's atmosphere.

15-37 (b) Gamma rays do not penetrate the Earth's atmosphere.

15-38 (b) Consider sources of very high energy within the galaxy.

15-39 Consider, for example, the implications of Exercise 15-28.

15-40 Consider the effect of the vast interstellar distances to be covered by any visitor.

15-41 (b) Consider the effect of commercial radio and television.

15-42 They are stars that are similar to the Sun.

15-43 Consider such things as the expense, the time spans involved, and the possible impact of success.

SOLUTIONS TO EXERCISES ON THE MILKY WAY GALAXY

15-1   (a)   It is disk-shaped with
             the following dimensions:

       (b)   Estimates run from 125
             to 300 billion stars.

       (c)   Common spiral galaxy.

15-2   (a)   Many other galaxies are spirals; the distribution of stars indicated
             its disk shape.

       (b)   By observations of bright emission nebulae which seemed to be dis-
             tributed along "arms" in the solar neighborhood.

       (c)   Neutral hydrogen is concentrated along the spiral arms, and it emits
             radio signals that are used to determine the spiral distribution of
             the gas and thus of the arms.

15-3   From their distance and direction from the Sun, Shapley determined that the
       globular star clusters occupy a spherical region in space. He assumed that
       the center of the sphere was the center of the galaxy, which put the Sun
       out in the galactic plane about 30,000 light years from the center.

15-4   (a)   The galaxy would be a sphere, with the Sun near the center.

       (b)   The galaxy would be a sphere, with the Sun near the edge; or the
             galaxy would be a disk, with the Sun above or below the center.

       (c)   The galaxy would be a disk, with the Sun near the edge.

15-5   (a)   The brighter stars are rather uniformly distributed all around the
             celestial sphere, with perhaps a slight concentration of stars toward
             the region of Orion.

       (b)   Same as (a), except there is a great concentration of faint stars along
             the band called the milky way, and the band is brightest in the direc-
             tion of the constellation Sagittarius.

       (c)   The faint, milky way stars cannot be resolved into individual stars with-
             out a telescope.

15-6   (a)   Sagitarrius.

       (b)   The interstellar dust blocks the light from the distant nucleus.

       (c)   The milky way.

15-7   (a)   Doppler shift in their spectra indicate that all objects in the universe
             rotate; in particular other galaxies rotate. Also, observations of the
             21-cm radiation from the hydrogen in the spiral arms indicate that the
             arms orbit the nucleus.

15-7   (b)   The Sun is in orbit about the nucleus of the galaxy, with a period
             of about 200 million years.

15-8
                                no doppler shift
                                  p.m. to left
          Stars receding   stars approaching
             no p.m.   SUN——→   no p.m.

                              no doppler shift
                                p.m. to right

                                        │   to galactic center
                                        ↓

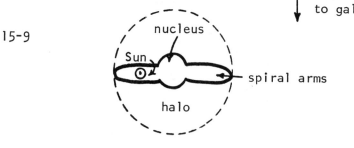

15-9

15-10 (a)   Interstellar gas and dust,   (b)   Globular clusters,
             galactic clusters,   Planetary nebulae,
             stellar associations,   Pop. II Cepheid variables,
             HI & HII regions,   RR Lyrae variables,
             Young, bright O & B stars,   Interstellar gas.
             Metal-rich Pop. I stars,
             Stars of all ages and types.

       (c)   Dense concentration of stars, probably Pop. II. Perhaps a few
             globular clusters.

15-11 Initially we studied the nucleus of the galaxy in Andromeda (M 31) and
       assumed that it is like our own. Now we also observe and study the radio
       emissions from the nucleus of our galaxy.

15-12 $M_{galaxy} = \dfrac{a^3}{P^2} = \dfrac{(2 \times 10^9 \text{ au})^3}{(2 \times 10^8 \text{ yr})^2} = 2 \times 10^{11}$ solar masses.

15-13 (a, b) The galaxy in Andromeda (M 31) is visible from the U.S.   Two of
               the companion galaxies of the Milky Way (large and small Magellanic
               Clouds) are visible from the southern hemisphere. (The third compan-
               ion galaxy is observable only with radio telescopes.)

15-14 (a)   Interstellar dust in the galactic plane obscures the light from the
             nucleus.

       (b)   Radio waves from the nucleus can be detected because they penetrate
             the interstellar dust.

15-15 (a)   The dark nebulae are interstellar dust clouds that obscure bright
             objects behind them. Also, the tenuous dust polarizes the starlight
             that does shine through, and it reddens the starlight since the shorter
             wavelengths are obscured more than the longer. It is also believed
             that dust plays an important role in star formation.

15-15 (b) In or near the galactic plane, in the spiral arms.

15-16 (a) A region of hot gas, glowing because there is one or more hot stars imbedded within or nearby.

(b) Often reddish, the characteristic color of glowing hydrogen.

(c) In the spiral arms.

(d) The Orion and Lagoon nebulae are well-known emission nebulae.

15-17 In the nebular environment the electron transitions of hydrogen may not produce much visible light; rather the transitions of the less abundant elements like oxygen and nitrogen may in fact emit more visible light.

15-18 (a) A region of gas and dust that is too far from a hot star to be heated to emission, but it is close enough to reflect the starlight.

(b) Usually bluish, because shorter wavelengths of light are reflected more than the longer wavelengths of light.

(c) In or near the spiral arms of the galaxy.

(d) The nebulosity around the stars in the Pleiades cluster.

15-19 (a) A cloud of dust which obscures the light from objects behind it.

(b) In or near the spiral arms.

(c) The Horsehead Nebula in Orion is well known.

15-20 Some of the dark nebular material might evolve into a star, and the star's energy would excite the nebula to glow.
A bright star might move into the region of the nebula, causing the nebular material to glow.

15-21 (a) Cool gases radiate in the radio part of the spectrum and are detected by radio telescopes. Hot gases radiate visible light and are observed by optical telescopes and spectrascopes.

(b) Gas is throughout the entire galaxy, with some concentration in the spiral arms.

15-22 (a, b) It is an emission nebula which is so hot that its hydrogen is ionized, and any dust that might have been there has been evaporated.

15-23 (a) The percentage of light elements (hydrogen and helium) seems to have gradually decreased from about 99 in the early galaxy to about 96 now, because the stars convert hydrogen and helium into heavier elements and then return the heavier material to the interstellar medium.

(b) Older (Pop II) stars, formed from the early lighter material, show only about 1% heavy elements in their spectra; whereas stars formed recently (Pop I) show greater percentages of heavier elements in their spectra.

15-24 Starlight becomes polarized when it travels through the tenuous interstellar dust grains which have been aligned by galactic magnetic fields.

15-25 The polarization of starlight can only be caused (so far as we know) by dust grains that have been aligned by magnetic fields.

15-26 (a) It is radiation in the radio range of the spectrum that is produced by electron transitions in neutral hydrogen atoms in space.

       (b) 21-cm radiation is detected by radio telescopes.

       (c) Its presence was first predicted from atomic theory in 1944; it was first observed in 1951.

       (d) Observations indicate the presence of neutral hydrogen, and doppler shift may indicate the radial velocity of the hydrogen.

15-27 These are spectral lines caused by electron transitions that do not occur on the Earth. They do occur in the extreme environment of space, however, so their "forbidden" spectra are observed in astronomical objects.

15-28 (a) CH_3OH methyl alcohol,   NCHO formaldehyde,

            HC_3N cyanoacetylene,   HCN hydrocyanic acid,

            NH_3 ammonia,   many others.

       (b) Their presence in the interstellar medium has been detected by radio telescopes, since each substance emits or absorbs radio signals at characteristic wavelengths.

15-29 (a) Dark nebular material forms the horse's head and the dark region below it. The light region around the head is glowing gas, and the white spots are stars.

       (b) Most of the stars are in the foreground; the emission nebula is in the background, obscured by the dust in the lower part of the photo.

15-30 Look in the nebular regions of the galaxy, because that is where the gas and dust are located. Stars are made from gas and dust.

15-31 (a) Observationally, Pop I objects have about 4% heavy elements; whereas Pop II have only about 1%. Pop I objects are thought to be relatively younger than Pop II objectives.

       (b) Pop I: open clusters, young stars, Type I Cepheids, supergiants and upper main-sequence stars.

            Pop II: Type II Cepheids, RR Lyrae, planetary nebulae, "high velocity" stars.

       (c) Pop I objects are associated with the spiral arms. Pop II objects are found throughout the entire galaxy, and are believed to be associated with the halo.

15-32 Older stars were formed at a time when the interstellar material was about 99% hydrogen and helium. Younger stars were formed later when the heavier (than helium) elements composed 4-5% of the interstellar medium.

15-33 (a, b) These are stars whose velocities relative to the Sun are unusually high. It is thought that they are not intrinsically fast-moving stars; rather their elongated galactic orbits are quite different from the Sun's near-circular orbit.

15-34 (a) Only those novae located relatively near to the Sun are observed. All others are obscured by the interstellar dust.

(b) Novae are believed to occur near the ends of the long lives of moderate mass stars. As time passes, more stars evolve to this stage.

15-35 (a) Young stars that are surrounded by nebular gas and dust, like Eta Carina. The center of the galaxy.

(b) Very hot stars, the Large Magellanic Cloud, the Sun.

(c) Infrared radiation is detected by telescopes operating in aircraft or in very dry regions at high altitude.

Ultraviolet radiation is detected by instruments carried above the atmosphere by rockets and satellites.

15-36 (a) Some X-ray sources are known to be remnants of supernovae, others are members of binary star systems--perhaps black holes. Most are as yet unidentified with optical sources.

(b) They lie near the plane of the galaxy, with some concentration toward the nucleus.

(c) From satellites or rockets sent above the Earth's atmosphere.

15-37 (a) A few supernovae remnants are known to emit gamma radiation.

(b) With special detectors flown on satellites.

15-38 (a) Primary cosmic rays are high-energy, high-speed particles (protons, helium nuclei, and heavier elements) which bombard the Earth's atmosphere.

(b) Origin is not definitely known; supernovae are strongly suspected.

15-39 Evidence of interstellar organic molecules has been found. At least one meteorite has been found on Earth that contained organic substances not native to Earth.

Several nearby stars are thought to have planetary sized companions.

Many stars have environments that are thought to be suitable for the evolution of life as we know it.

15-40 Interstellar distances are so great as to make travel to our solar system very unlikely, even from a nearby star.

15-41 (a) Regular broadcasts of radio signals would send the message at the speed of light to anyone who happened to be listening.

15-41 (b) For about 30 years the Earth has been emitting large amounts of energy in TV broadcasts. This has now reached the stars within 30 light years of the Sun.

15-42 Main sequence stars of spectral classes F, G and K are thought to be the best candidates, because they have regions around them ("habitable zones") in which the environments could have permitted life to evolve.

15-43 It could be very expensive, and there is no guarantee of success regardless of the amount spent.

Success, if achieved, might not occur for decades or centuries, even on a program begun today.

The impact of success could cause unforseen psychological problems.

The attempt might unite a devisive nation (or nations) in a common goal. Success might unite the entire globe of nations.

Success could be rewarded by access to the vast store of knowledge accumulated by advanced civilizations--our "galactic heritage."

Appearances and Types

16-1 (a) What is a galaxy?

(b) Why do galaxies appear so small on the celestial sphere?

16-2 (a) Name the three principal types of galaxies, and describe their main features.

(b) Approximately what percentage of all galaxies fall into each of the three types?

16-3 Why do galaxies have different appearances, masses and sizes?

16-4 (a) What type of galaxies resemble globular clusters, and in what ways are they similar?

(b) How do we know that those galaxies are not, in fact, globular clusters that were ejected from spiral galaxies?

16-5 At one time it was thought that galaxies evolved from one type to another as they grew older.

(a) Describe the two schemes of galactic evolution that have been proposed in the past.

(b) What evidence leads us to now believe that neither scheme is correct?

16-6 (a) Why is it suggested that giant elliptical galaxies are quite old, and spirals (like the Milky Way) are younger?

(b) Cite one piece of evidence which suggests that spiral galaxies are not so young.

16-7 Explain what is meant by the following designations for galaxies.

(a) NGC 4151,
(b) M31,
(c) 3C405,
(d) Centaurus A.

16-8 How did the term "spiral nebula" come into usage, and why is it no longer used?

16-9   (a) How was it determined that the so-called "spiral nebulae" were in fact external galaxies, and not part of the Milky Way?

       (b) Why wasn't this determination made before the 1920's?

16-10 How does one distinguish between the images of stars and distant galaxies on a photograph?

16-11 (a) What presently limits the faintness to which galaxies can be photographed?

       (b) How might this limit someday be extended?

16-12

Exercise 16-12. Messier object 104. (Hale Observatories photo.)

       (a) What kind of object is shown in the photo?

       (b) What is producing the light seen in the center of the object?

       (c) What is causing the dark band across the center of the object?

16-13 What evidence indicates that some galaxies may be much larger (5 to 10 times) than their visual images?

Exercise 16-14   The galaxy NGC 4151.   (Hale Observatories photo.)

16-14   (a)   Identify the types of galaxies that appear in the center and in the lower left of the photo.

(b)   What are the distinguishing features of the two types of galaxies?

The Local Group of Galaxies

16-15   (a)   What is the Local Group of galaxies, and about how many members does it have?

(b)   What are the largest galaxies in the Local Group, and what type are they?

(c)   What is the most common type of galaxy in the Local Group?

(d)   Where might undiscovered members of the Local Group be located?

16-16   Describe the two well-known satellite galaxies of the Milky Way--their distance, type, and general form.

16-17 Another satellite galaxy of the Milky Way was identified in 1975, and humorously named "Snickers."

(a) How was the new satellite galaxy discovered?

(b) How far away and in what direction from the Sun is it, and how does this distance compare with the distances to the Magellanic Clouds?

(c) Why wasn't this galaxy discovered before 1975?

16-18 What is the relationship between the constellation of Andromeda and the galaxy M31 in Andromeda?

16-19 How does the study of the spiral galaxy M31 contribute to our knowledge of our own Milky Way galaxy?

16-20 The nearby galaxy M31 is approaching the Milky Way at a speed estimated to be about 300 km/sec.

(a) How is this known?

(b) Under what condition would the two galaxies collide?

(c) When would this collision occur, if it happened?

(d) Describe what would happen to the members (the stars) of the Milky Way during and after such a collision.

*16-21 Is it possible to photograph stars of the Sun's brightness within the galaxy M31, using the 200-inch Hale telescope? Ignore the effect of interstellar absorption.

Galactic Distances

16-22 (a) State in words some ways to estimate the distance to a nearby galaxy.

(b) Which of these methods may be used for distant galaxies?

*16-23 Suppose a cepheid variable is observed in the spiral arms of a distant galaxy. Its average apparent magnitude is +20, and its period is 30 days. Calculate the distance to the galaxy.

*16-24 (a) Calculate the distance to a galaxy in which is observed a Type I supernova. At maximum brightness the supernova's apparent magnitude is observed to be +15.

(b) If the interstellar medium has dimmed the supernova's light by two magnitudes, is your calculated distance too large or too small?

16-25 (a) Why is it believed that the doppler shift in the light from a galaxy is related to its distance?

(b) Cite two other possible causes for the red-shift in the light from galaxies.

(c) Cite reasons why the two possible causes mentioned in (b) are rejected by most astronomers.

16-26 (a) State in words what is the Hubble constant.

(b) State in algebraic symbols what is the Hubble constant.

(c) What is the currently accepted value of the Hubble constant?

16-27 (a) Define the parameter z , used in the Hubble relationship.

(b) What is the range of values of z that have been observed for galaxies and for quasars?

(c) Assuming the Hubble constant is 70 km/sec/Mpc, what maximum distances do the values in (b) correspond to?

16-28 Calculate the distance to a galaxy in which the K line of Calcium is observed at a wavelength of 4760 A.

16-29 In what way do astronomers look into the past?

16-30 Why can't astronomers observe the parallax and proper motions of stars in galaxies outside the Milky Way?

Energy Output of Galaxies

16-31 Cite some examples of violent events that appear to be occurring in and around extra-galactic objects.

16-32 Compare the energy outputs of the following

(a) The Sun,
(b) An ordinary galaxy with 10^{11} Suns,
(c) A supernova,
(d) A quasar,
(e) An energetic radio galaxy.

16-33 Describe the two or three brightest extra-galactic radio sources.

16-34 Some galaxies show twin radio sources located symmetrically on each side of the optical image. What is a possible explanation for the twin radio sources?

Groups and Clusters of Galaxies

16-35 (a) What evidence indicates that two or more galaxies are physically related and probably orbiting each other?

(b) Name one system of orbiting galaxies within the Local Group, and one system outside the Local Group.

16-36 (a) What evidence indicates that galaxies tend to cluster together?

(b) Name two clusters of galaxies, and tell why they have those particular names.

16-37 (a) How many galaxies are found in a typical cluster, and in a rich cluster?

(b) Is the Local Group an ordinary or rich cluster?

16-38 What is a major question presented by the clusters of galaxies?

Quasi-Stellar Objects (Quasars)

16-39 Describe the basic observed features of quasars.

16-40 Cite two ways that quasars have been discovered.

16-41 What is the basic problem with the quasars?

16-42 (a) Describe the problem of relating the distance, redshift, and energy output of a quasar.

(b) What evidence indicates that quasars are not members of the Milky Way galaxy?

16-43 Cite some explanations or models which have been offered for the quasars. For each, cite one contradictory concept or observation.

16-44 What evidence suggests that the Hubble relationship (distance vs redshift) may not be the same for quasars as it is for galaxies?

16-45 What evidence suggests that the quasars are galaxies in early stages of evolution?

*16-46 Estimate the radius (in light years and in km) of a quasar with a period of variability of 50 days.

HINTS TO EXERCISES ON GALAXIES AND QUASARS

16-1   (b) Consider their vast distances.

16-2   The types are spiral, elliptical, and irregular.

16-3   For many questions in astronomy the answers are not yet known.

16-4   (a) Compare them in shape and appearance.

       (b) Compare the radial velocities of galaxies with those expected
            from ejected objects.

16-5   See Hubble's "tuning fork" diagram.

16-6   (a) This idea arises from the amounts of gas and dust in the galaxies.

       (b) Consider the ages of globular clusters.

16-7   The first three refer to catalog designations.

16-8   The term arises from the understanding of the nature of galaxies prior
       to the 1920's.

16-9   See Exercise 13-12.

16-10 Look carefully at a photo of a cluster of galaxies.

16-11 Even with the largest telescope the astronomer must look through the
       Earth's atmosphere.

16-12 The object is several million light years from the Sun.

16-13 Galaxies emit radio emission as well as visible light.

16-14 Among the types of galaxies are regular spiral, barred spiral, irregular,
       elliptical, dwarf, and Seyfert.

16-15 Consider the regions where members have been recently discovered. Also,
       see Exercise 12-22(b).

16-16 They are both irregular galaxies.

16-17 (a) It was discovered by analysis of radio observations.

       (b) See Exercise 16-15.

16-18 Consider the directions in which they are both observed.

16-19 See Exercise 15-11.

16-20 (a) This comes from analysis of the spectrum.

       (c) The distance of M31 is about 1.9×10^{19} km.

16-21   The limiting photographic magnitude of the 200 inch is about +24. The distance of M31 is about 600,000 pc.

16-22   One method uses variable stars, and another uses the red shift in the galaxy's spectrum. The others utilize bright objects within the galaxy.

16-23   Assume the cepheid is Pop. I, and see Exercise 13-13.

16-24   Assume that at maximum brightness the absolute magnitude of the supernova is about -20.

16-25   (b) Consider the effect of light leaving a strong gravitational field.

16-26   It relates the distances and observed redshifts of galaxies.

16-27   See Exercise 16-26 (b).

16-28   Calcium at rest emits light at 3967 $\overset{\circ}{A}$. See Exercises 3-28 and 16-27.

16-29   Consider the vast distances of the galaxies, and the time required for light to reach us from a galaxy.

16-30   Consider the distances to the galaxies. See Exercise 11-9.

16-31   Consider exploding galaxies and some quasars.

16-32   The solar energy production is given in Exercise 10-40. Assume the galaxy produces 10^{11} times as much energy as the Sun.

16-33   They are galaxies.

16-34   It is thought that violent events have ejected material from these galaxies.

16-35   The evidence is seen on photographs of the galaxies.

16-36   Consider their appearances on photographs, and their redshifts.

16-37   (b) See Exercise 16-15 (a).

16-38   The problem is with the force required to hold the galaxies together as a cluster.

16-39   The observations involve spectra, redshift, color and radio emission.

16-40   Consider their unique spectral characteristics, and Exercise 16-39.

16-41   It has to do with their identity.

16-42   Quasars show no proper motions (b).

16-43   We would like the model to not only fit all the observations, but also to be consistent with the known laws of nature.

16-44   If objects appear very close together in the sky, it is probable that they are close together in space too.

16-45   Consider the vast distances at which we may be seeing the quasars, and the epoch in the history of the universe during which they occurred.

16-46   The speed of light is $c = 3 \times 10^5$ km/sec.

16-1 (a) Groups of millions and billions of stars, located outside our own Milky Way galaxy.

(b) They appear small because they are at vast distances from us.

16-2 (a) Irregular galaxies (about 3%) have no definite form or shape.

(b) Spiral galaxies (about 25-30%) consist of a nucleus, a disk with spiral arms, and a halo or corona. Most have a great deal of interstellar matter in their disk regions.

(c) Elliptical galaxies (about 70-75%) are ellipsoidal in shape, with no spiral arms and little interstellar matter.

16-3 The reason is not yet known.

16-4 (a) The small spherical E0 galaxies resemble globular clusters in appearance and in the kinds of stars (old) they contain.

(b) If those small E0 galaxies had been ejected from large spirals, then we should observe many of them to be approaching us (blue-shifted spectra), but in fact most are receding (red-shifted).

16-5 (a) An early scheme is illustrated by Hubble's "tuning fork" diagram, which suggests that galaxies begin as ellipticals, evolve through spirals, and end as irregulars. Toward the middle of this century it was suggested that the evolutionary sequence was essentially the opposite.

(b) The galactic evolutionary sequence (if there is one) is uncertain because many galaxies seem to have both old and young objects within.

16-6 (a) Giant elliptical galaxies contain very little gas and dust; whereas spiral galaxies have large quantities of this material. Some astronomers believe that the amount of gas and dust in a galaxy decreases in time as this material gets locked in stars.

(b) The fact that relatively young spiral galaxies have very old globular clusters in them seems to be an inconsistency.

16-7 (a) The 4151st entry in the New General Catalog.

(b) The 31st entry in Messier's catalog of "nebulae."

(c) The 405th entry in the Third Cambridge Catalog of radio sources.

(d) The first radio source to be observed in the direction of the constellation of Centaurus.

16-8   Before their true nature was known, the external galaxies were called
        spiral nebulae because (a) many of them have spiral appearances, and (b)
        they were indefinite objects. The term is no longer needed because we
        now know the objects are galaxies.

16-9   (a) Some cepheid variable stars were photographed in the nearby galaxy
             M31. Their distance was determined, and it was found that the stars
             and the galaxy in which they were located were much farther away than
             the bounds of the Milky Way.

        (b) The cepheids appear so faint at the distance of M31, that their detec-
             tion had to await the construction of a very large telescope--the
             100-inch at Mt. Wilson Observatory in 1924.

16-10   Stellar images are round with sharp edges and uniform brightness. The
        brighter images have spikes or points around them. Galaxy images can be
        round, elongated or irregular. They are generally brighter in the center
        and somewhat fuzzy near the edges.

16-11   (a) The glow of the night-time atmosphere limits exposure times to about
             5 hours on the largest telescopes.

        (b) Use a large telescope above the Earth's atmosphere (in Earth orbit)
             or on the Moon.

16-12   (a) Spiral galaxy.

        (b) Light from stars and emission nebulae.

        (c) Interstellar dust.

16-13   Radio emission from the apparently empty region surrounding a galaxy
        would indicate the presence of matter which is part of the galaxy
        extended into that region.

16-14   (a) Center: Seyfert; lower left: barred spiral.

        (b) The Seyfert has an especially large, bright nucleus; the nucleus
             of the barred spiral is a bar from which the spiral arms trail off
             the ends.

16-15   (a) It is the group of about 20 galaxies within approximately 2 million
             light years of the Milky Way.

        (b) The two spirals: M31 and the Milky Way.

        (c) Most of the known members are dwarf elliptical galaxies.

        (d) They might lie near the plane of the Milky Way (in the so-called
             "zone of avoidance") or behind other members.

16-16   The Large and Small Magellanic Clouds are small irregular galaxies about
        170,000 LY from the Milky Way. They are chaotic without regular form,
        and contain stars, clusters, gas and dust.

16-17   (a) It was discovered as an anomaly on the radio maps of the hydrogen
             in the region of Gemini.

        (b) It is estimated to be about 55,000 LY from the Sun.

16-17 (c) It is hidden from visual observation by the interstellar material
in the Milky Way.

16-18 Both lie in the same direction, on the same part of the celestial sphere.
They are in no way physically associated, however, since the constellation
stars are members of the Milky Way and M31 is another galaxy.

16-19 M31 is relatively close to the Milky Way, and the two galaxies are similar
in size and structure. Hence, our view of M31 is probably similar to an
external view of the Milky Way. Also, we can study the nucleus of M31;
whereas we can't visually observe the nucleus of the Milky Way.

16-20 (a) From the average doppler shift (toward the blue) in the light from M31.

(b) If M31 has no proper motion.

(c) Time = distance/velocity = 1.9×10^{19} km$/300$ km/sec

$$= 6.3 \times 10^{16} \text{ sec} = 2 \text{ billion years.}$$

(d) Few star collisions would occur. The galaxies would pass through each
other with little effect, except that much of the gas and dust would
be swept out and left in space.

16-21 The absolute magnitude of the faintest observable star in M31 is found from
$\log d = (m + 5 - M)/5$, as

$M = m + 5 - 5 \log d = +24 + 5 - 5 \log (600{,}000) = 0.1$

Thus a star must be brighter than $M = 0.1$ to be photographed at that
distance. Stars like the Sun $(M = +5)$ are too faint.

16-22 (a) Use the period-luminosity relationship to estimate the distances to
some cepheids in the galaxy (see Exercise 14-15).

Assume an average absolute magnitude for some class of bright object
(HII region, globular cluster, 5 brightest blue stars, etc.), then observe their apparent magnitudes in the galaxy, and calculate the distance from $\log d = (m + 5 - M)/5$.

For a given type of galaxy (e.g., dwarf spiral) assume it has a particular diameter. Then the distance is estimated from the observed
angular diameter.

Use the Hubble relationship, after observing the red-shift.

(b) The last two methods can be used for distant galaxies.

16-23 From the period-luminosity curve, we find $M = -5$ for a period of 30 days.

Then $\log d = (m + 5 - M)/5 = \left[+20 + 5 - (-5)\right]/5 = 6.$

$$d = 10^6 \text{ pc} = 1{,}000{,}000 \text{ pc.}$$

16-24 (a) $\log d = \left[+15 + 5 - (-20)\right]/5 = 8$
$$d = 10^8 \text{ pc.}$$

(b) The calculated distance is too large.

16-25 (a) All other indicators of galactic distances show that the farther the galaxy the greater the red-shift in its spectrum.

(b) It is known that light is slightly red-shifted as it leaves the gravitational field of the galaxy, and it has been suggested (without proof) that light may lose energy from traveling long distances.

(c) The gravitational red-shift is independent of the distance to the galaxy--it should be the same for close and distant galaxies. The "tired light" hypothesis has no proof.

16-26 (a) It is the proportionality constant between the red-shift z of a galaxy and its distance d.

(b) Algebraically it is expressed by $z = Hd/c$, where H is the Hubble constant and c is the speed of light.

(c) Its value is thought to be somewhere between 100 and 50 km/sec per megaparsec.

16-27 (a) $z = \Delta\lambda/\lambda_0$, where λ_0 is the wavelength of the light emitted from a source at rest, and $\Delta\lambda$ is the observed change in the wavelength of the same light.

(b) Galaxies: 0 to about 0.4, Quasars: 0.15 to about 3.5.

(c) $d\ (z=0.4) = \dfrac{0.4 \times 3 \times 10^5 \text{ km/sec}}{70 \text{ km/sec/Mpc}}$ = 1.7 billion pc.

$d\ (z=3.5) = 3.5 \times 3 \times 10^5\ /\ 70$ = 15 billion pc.

16-28
$$\frac{\lambda - \lambda_0}{\lambda_0} = \frac{v}{c} = z = \frac{4760 - 3967}{3967} = 0.2$$

$$d = \frac{zc}{H} = \frac{0.2 \times 3 \times 10^5}{70} = 850{,}000 \text{ pc.}$$

16-29 When an object is observed, it is seen as it appeared when the light we receive left the object. The galaxy M31, for example, is seen as it appeared about 2 million years ago, since it takes about 2 million years for light to travel from M31 to us.

16-30 The galaxies are so far away that these motions are too small to detect.

16-31 The peculiar galaxies M82 and NGC 5128 (radio source Centaurus A) seem to be exploding. Also, some galaxies (e.g., M87) and quasars (3C273) appear to have ejected jets of material.

16-32 Sun 4×10^{33} ergs/sec

Galaxy approximately 4×10^{43} ergs/sec

Supernova 10^{49} to 10^{51} ergs/sec

Radio galaxy 10^{56} – 10^{61} ergs/sec

Quasar approximately 10^{58} ergs/sec

16-33 One of the brightest sources is the double galaxy Cygnus A. Another is what appears to be an exploding galaxy Centaurus A.

16-34 One theory proposes that the two radio sources are clouds of hot gas that were ejected from the galaxy in opposite directions.

16-35 (a) Bridges of luminous material are seen to extend between some galaxies. Also, galaxies that appear close together in the sky are found to have nearly identical red-shifts.

(b) The Milky Way and its three satellites are an example in the Local Group. Stephan's Quintet is a distant example.

16-36 (a) On photographs, distant galaxies appear to clump together. Also, the galaxies in what appears to be a cluster all have about the same red-shifts.

(b) Two of the better known are the Coma and Virgo clusters, named for the constellations in which they appear.

16-37 (a) From 5 to 20 in an ordinary cluster, hundreds in a rich cluster.

(b) Ordinary cluster.

16-38 Cluster members should be gravitationally bound to the group, or else the cluster would dissipate in time. The observable galaxies in a cluster, however, seem to contain only about 10% of the mass needed to hold the cluster together. At present we cannot account for this "missing mass."

16-39 Their spectra are quite different from spectra of stars and nebulae. The continuum is quite bright in blue and infrared. They show wide emission lines and often absorption lines. The spectral lines are shifted far to the red, and one spectrum may contain several red-shifts. Some emit large amounts of radio energy. They have star-like appearances.

16-40 Radio quasars are first identified as radio sources. Then the radio source is correlated with a bluish star-like image, and then the spectrum is found to be that of a quasar.

16-41 Astronomers are unable to develop a model which is consistent with all the observations and with the known laws of nature.

16-42 (a) If they are at the vast distances indicated by their red-shifts, then they must (by some unknown means) emit more energy than entire galaxies. If they are close, on the other hand, then some unknown process is responsible for the large red-shifts.

(b) They show no proper motions, and there is no accepted explanation for such large red-shifts in members of the Milky Way.

16-43   Distant early galaxies: Unknown source of energy, and a smaller size is
         indicated by the period of variability.

         Objects ejected from the Milky Way or the Local Group: Implies unique-
         ness of man's region of the universe.

         Supermassive stars: Theory and local observations indicate that stars
         cannot exist at much more than 100 solar masses.

         Objects ejected from other galaxies: No blue shifts.

16-44   There are several cases where a quasar and a galaxy appear very close
         together in the sky (within a few arc minutes) yet have very different
         red-shifts.

16-45   Their red-shifts indicate that many quasars are more distant than the
         farthest galaxies. The farther away we look, the farther back in time we
         are seeing. Hence we may be seeing the quasars as samples of the universe
         at an early epoch, when galaxies were just forming.

16-46   50 days = T = 4,320,000 seconds.

         Radius $R = T \times c = 4,320,000 \text{ sec} \times 3 \times 10^5$ km/sec,

                  $= 13 \times 10^{11}$ km = 0.14 light years.

General Exercises

17-1   What are the three basic "ground rules" which any scientific hypothesis (including cosmological models) should follow?

17-2   What is the difference between cosmology and cosmogony?

17-3   (a) What is the cosmological principle?

        (b) What is the perfect cosmological principle?

        (c) What observational evidence (if any) supports or contradicts the principles of parts (a) and (b)?

17-4   If the quasars are in fact at cosmological distances, what would that say about the validity of the perfect cosmological principle?

17-5   (a) What is the 3° background radiation?

        (b) How does the existence of this radiation seem to support the evolutionary or big bang cosmological model?

17-6   Why is it so difficult to determine the location of the center of the universe?

17-7   It has been suggested that approximately half of the material in the universe should be composed of antimatter.

        (a) What is antimatter?

        (b) What is the basic concept which leads to the above suggestion?

        (c) Why is it difficult to observationally test the suggestion?

17-8   (a) Explain how a line, which is perceived to be one-dimensional, straight, and to go to infinity in both directions, might actually be a closed two-dimensional figure.

        (b) Explain how a plane, which is perceived to be flat, two-dimensional, and to go to infinity in both dimensions, might actually be part of a closed three-dimensional figure.

17-9 (a) What do observations indicate about the relative motions of galaxies in the universe, on a large scale?

       (b) What is the observation or evidence from which the answer to part (a) is determined?

17-10 If the galaxies are all moving away from each other, are they moving at constant speed, are they accelerating, or are they slowing down? Explain your answer.

17-11 Explain how it is possible for all the galaxies to be moving away, with the farthest moving the fastest, without the Milky Way galaxy being in the center of the universe.

17-12 If the universe is expanding and the galaxies are moving away from us, how do we explain those few galaxies with spectra that are blue-shifted?

17-13 Some are tempted to explain the red-shift in galactic spectra as due to absorption of the light by intergalactic dust. Why is this an incorrect explanation?

17-14 A small cluster of galaxies is found to have a radial velocity of 35,000 km/sec, moving away from the Milky Way galaxy.

       (a) At what percentage of the speed of light is the cluster moving?

       (b) Calculate the distance to the cluster assuming the Hubble constant H = 70 km/sec per million parsecs.

       (c) How does the distance compare to those of the most distant objects known?

*17-15 Calculate an upper limit for the age of the universe according to the big bang cosmological model. Assume the Hubble constant H = 70 km/sec per million parsecs.

17-16 What is meant by the problem of the missing mass?

17-17 It has been suggested that the missing mass, which would comprise about 90% of the universe, might be in black holes.

       (a) Cite an argument in favor of this hypothesis.

       (b) Cite an argument in contradiction to this hypothesis.

17-18 (a) Describe some ways to test whether the expansion of the universe will ever stop--to see if the universe is open or closed.

       (b) What are the results of these tests to date?

Cosmological Models

17-19 (a) What are the basic features of the big bang model?

(b) What evidence supports this model?

(c) What evidence contradicts this model?

17-20 (a) What are the main features of the steady state model?

(b) What evidence supports the steady state model?

(c) What evidence contradicts the steady state model?

17-21 Describe the beginning and the end of the universe as proposed by each of the following cosmological models:

(a) Steady state,

(b) Big bang - closed,

(c) Big bang - open.

17-22 In the big bang cosmological model,

(a) What are the two possible ends of the universe as we know it?

(b) What property or characteristic of the universe will determine which of the ends will occur?

(c) Toward which of the two ends does the evidence seem to point?

17-23 In what ways do the big bang cosmological models conflict with western theology?

17-1   These have to do with observations, the laws of nature, and man's place in the universe.

17-2   Consult a dictionary or glossary of your text.

17-3   They describe the variations to be found throughout the universe.

17-4   The farther away an object is located, the farther back in time we look to observe it.

17-5   It is analogous to radiation that would be observed from a black body at three degrees above absolute zero.

17-6   Consider the definition of the word center.

17-7   Antimatter differs from matter only in the electrical charges carried by the electron and proton.

17-8   Consider imperceptible curvatures in the line and plane.

17-9   See Exercises 16-25 and 16-26.

17-10   This is determined from Newton's law of gravitation.

17-11   Suppose we can observe only a small part of the universe.

17-12   Consider the motions of individual galaxies within clusters of galaxies.

17-13   It is known that dust in the galaxy absorbs starlight, and that the shorter wavelengths are absorbed more than the longer (red) wavelengths.

17-14   Radial velocity = Hubble constant x distance. Algebraically,
        V_r (km/sec) = H (km/sec per million parsecs) x d (parsecs).

17-15   There are about 3×10^7 seconds/year, and 3×10^{13} km/parsec.

17-16   See Exercises 16-38 and 17-22 (b).

17-17   See Exercise 15-34 for the properties of black holes.

17-18   (b) See Exercise 17-22 (c).

17-19   This model is in contradiction with the perfect cosmological principle.

17-20   The steady state model derives from the perfect cosmological principle.

17-21   Some models propose no specific beginning or end.

17-22   (a) Both involve the consequences of the expansion.

17-23   Remember that cosmology only attempts to explain what matter has been doing since it existed in the universe.

SOLUTIONS TO EXERCISES ON COSMOLOGY

17-1   The hypothesis should be consistent with the observed universe.
        The hypothesis should be consistent with the known laws of nature.
        The hypothesis should not require that man have a special place (in time
        or in space) in the universe.

17-2   Cosmology is the study of the organization and evolution of the universe;
        whereas cosmogony confines itself to the study of the origin of the
        universe.

17-3   (a, b) The principle states that the universe is homogenous throughout--
                basically the same everywhere in space, on a large scale. The per-
                fect cosmological principle is the same, but in addition states that
                the universe is the same for all time, past, present and future.

        (c) Observations of galaxies seem to confirm the perfect cosmological
             principle; however, recent observations of quasars may show that the
             universe was very different in the distant past.

17-4   If the quasars are very distant--many of them farther than the most distant
        galaxies--then we are seeing objects (quasars) which existed only a long
        time ago since no quasars are seen nearby. Thus the universe was very
        different in the distant past, and the perfect cosmological principle is
        in error.

17-5   (a) It is electromagnetic radiation (photons) that appears to come from
             all parts of the sky with equal intensity. Its wavelength and fre-
             quency characteristics are the same as would be observed of a black
             body at a temperature of 3° K.

        (b) Calculations show that if the universe began with a "big bang" several
             billion years ago, then some of the photons that were produced by the
             initial explosion should persist today in basically the form of the
             observed 3° background radiation.

17-6   The center is by definition equidistant from all edges. But we have no
        idea where the edges of the universe are , or even if it has edges.

17-7   (a) In antimatter the electric charges are reversed--the electrons are
             negative and the protons are positive.

        (b) There is no known reason why matter should exist with only positive
             protons and negative electrons. Also, antimatter has been observed
             in small quantities on Earth.

        (c) From a distance antimatter appears to behave just like matter, so
             there is no way to determine which a distant object is.

17-8   (a) The line might be a small segment of a giant hoop, whose curvature
             is too small to measure locally.

17-8   (b) The plane might be a small portion of a large sphere, whose curvature is too small to measure locally.

17-9   Observations of the red-shift in the spectra of galaxies indicate that the galaxies are moving away from each other, and that the greater the distance between galaxies the faster they move apart. In other words, the volume of space occupied by the galaxies is expanding.

17-10 The galactic radial velocities must be slowing down, because the mutual gravitational attractions of the galaxies for each other would tend to pull them together and resist the motion of expansion.

17-11 It is possible, for example, that the observable universe is only a small part of the entire universe. While the observable universe is expanding, that same observable universe could be moving away from (or toward) a center that we don't even see.

17-12 Although an entire cluster of galaxies is moving away from us, the random motions of the cluster members can result in a few of them approaching us at any given instant.

17-13 Intergalactic dust would absorb shorter wavelengths of light, causing what light does reach us to appear reddened; whereas we observe that all the light is shifted toward the red, but none is absorbed. Also, there is no other evidence (such as polarization) of intergalactic dust.

17-14 (a) $\frac{35,000}{300,000}$ x 100% = 12% of the speed of light.

       (b) $d = \frac{Vr}{H} = \frac{35,000 \text{ km/sec}}{70 \text{ km/sec per } 10^6 \text{ pc}}$ = 500 x 10^6 pc = 500 million pc.

       (c) Some quasars are believed to be billions of parsecs away.

17-15 Age = $\frac{1}{H}$ = $\frac{1 \text{ million parsecs}}{70 \text{ km/sec}}$ x 3 x 10^{13} km/pc = 4.3 x 10^{17} sec.

       4.3 x 10^{17} sec x $\frac{1 \text{ year}}{3 \times 10^7 \text{ sec}}$ = 1.4 x 10^{10} yr = 14 billion yrs.

17-16 There are two such problems which may be related. One involves the mass apparently needed to retain members of clusters of galaxies, and it is discussed in Exercise 16-36. The other is that if the universe is closed, then additional mass must be found. There is insufficient observable mass in the galaxies to stop the present expansion.

17-17 (a) If most of the early universe was in massive stars, then those stars would have evolved to invisible black holes by now.

       (b) If so much of the mass of the universe was in very massive stars, then some of the high-energy photons which those stars produced should be in evidence today, but they are not.

17-18 (a) Compare the ages of objects in the universe with the age given by the
            Hubble constant (Exercise 17-15).

            Examine the red-shifts of very distant objects to see if they are
            slowing down enough so that they will eventually stop.

            Observe directly the matter in the universe through counts of galaxies
            and clusters of galaxies, and see if there is enough to close the universe.

       (b) Current tests indicate that the universe is open--that there is only
            about 10% of the mass needed to stop the expansion.

17-19 (a) In the beginning all matter in the universe was together in one place--
            the ylem. An explosion ejected the matter outward at varying speeds.
            As the matter cooled hydrogen and helium were formed. Later the
            material began to clump together under gravitation, and galaxies were
            formed.

       (b) The red-shifts in galactic spectra and the 3° background radiation
            both support the model.

       (c) No serious contradictions are known; however it is not exactly known
            how galaxies would form from the material ejected by the primeval
            fireball.

17-20 (a) On a grand scale the universe is unchanging, both in time and space.
            Fresh new hydrogen is continuously created to maintain a constant
            density of matter in the universe as the galaxies move away from each
            other.

       (b) The model is unsupported by the observed uniform density of galaxies
            throughout the universe.

       (c) The model is contradicted by the observations of quasars if they are
            at great distances, and by the 3° background radiation.

17-21 (a) No specific beginning or end.

       (b) The universe that we know began when all matter was in the ylem. It
            is not certain what will happen when the matter falls back upon itself
            after the expansion stops.

       (c) The universe began when all matter was in the ylem. It has no
            definite end.

17-22 (a) Open universe: The expansion continues forever.
            Closed universe: The expansion eventually stops, and then the galaxies
            move toward each other toward an uncertain fate.

       (b) The amount of mass in the universe.

       (c) Contemporary evidence points toward an open universe--insufficient
            mass to stop the expansion.

17-23   In my opinion there is no basic conflict. Cosmology, cosmogony, and science in general attempt to describe the behavior of matter for as long as it has existed in the universe. Modern science makes no attempt to explain <u>how</u> or <u>why</u> the matter came into existance. That, in my opinion, is the realm of theology, upon which science makes no encroachment.

17-23   In my opinion there is no basic conflict. Cosmology, cosmogony, and science in general attempt to describe the behavior of matter for as long as it has existed in the universe. Modern science makes no attempt to explain <u>how</u> or <u>why</u> the matter came into existance. That, in my opinion, is the realm of theology, upon which science makes no encroachment.